TeeJay

Maths

CfE First Level
Book 1B

Tom Strang, James Geddes, James Cairns, Dr Naomi Norman and Catherine Murphy

Orders: please contact Hachette UK Distribution, Hely Hutchinson Centre, Milton Road, Didcot, Oxfordshire, OX11 7HH. Telephone: +44 (0)1235 827827. Email education@hachette.co.uk Lines are open from 9 a.m. to 5 p.m., Monday to Friday. You can also order through our website: www.hoddereducation.co.uk. If you have queries or questions that aren't about an order, you can contact us at hoddergibson@hodder.co.uk

© Tom Strang, James Geddes, James Cairns 2023

Special thanks to Dr Naomi Norman and Catherine Murphy for their significant contribution.

First published in 2023 by

TeeJay Publishers, an imprint of Hodder Gibson, which is part of the Hodder Education Group

An Hachette UK Company

50 Frederick Street

Edinburgh, EH2 1EX

Impression number	5	4	3	2	1
Year	2027	2026	2025	2024	2023

Cover illustration © Ai Higaki/D'Avila Illustration Agency

Illustrations by Aptara, Inc.

Typeset in FS Albert 18/20 pts by Aptara, Inc.

Produced by DZS Grafik, Printed in Slovenia

A catalogue record for this title is available from the British Library.

ISBN: 978 1 3983 6323 6

SCOTLAND EXCEL

We are an approved supplier on the Scotland Excel framework.

Find us on your school's procurement system as

Hachette UK Distribution Ltd or *Hodder & Stoughton Limited t/a Hodder Education.*

MIX
Paper | Supporting responsible forestry
FSC www.fsc.org FSC™ C104740

Contents

Introduction
How will this book help me learn maths?

This book begins with a Chapter 0. This chapter is full of questions to help you revise maths you already know.

Each chapter in this book is packed with lots of practice covering all the maths you need to learn.

Each topic begins with a short explanation to get you started. Sometimes there are also examples to help you understand the ideas before you answer the questions in the exercise.

Review everything you have learnt in the end-of-year chapter at the end of the book.

Remember, remember

Sometimes you need to remember some maths that you have learnt before, this is in these boxes.

Let's try this!

These boxes include play-based activities that let you learn while you have fun. Play them with your classmates, with a partner or a friend.

Revisit, review, revise

Use these questions at the end of each chapter to look back at what you have learned.

This tells you what you're going to learn in each chapter.

Clear examples help you understand the ideas.

Exercises help you practise the maths.

These yellow boxes introduce you to some new maths.

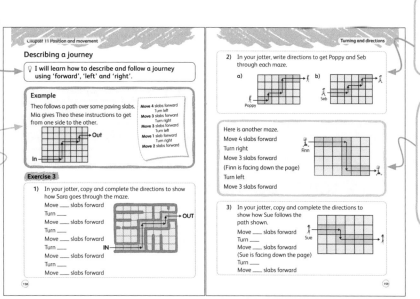

0 Revision of Book 1A

1) How many sweets altogether in each pair of jars?

a)

b) 

2) What numbers do the arrows point to?

a)

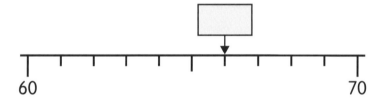

b)

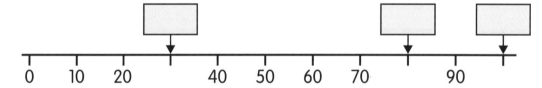

3) In your jotter, write all the **missing** numbers.

a)   4, 5, ___, 7, ___, 9, ___, 11, 12, ___

b)   35, 34, ___, 32, 31, ___, ___, 28, 27, ___

4) 23 means 2 **tens** and 3 **ones**.
In your jotter, copy and complete:

a)   17 means 1 **ten** and ___ **ones**.

b)   32 means ___ **tens** and ___ **ones**.

5) Copy and complete in your jotter:

   a) 67 = 60 + ____

   b) 95 = ____ + 5

   c) 36 = ____ + ____

6) Copy and complete in your jotter.
You may use counters to help you.

   a) 14 + 4 =

   b) 53 + 6 =

7) Mario's sister is **9** years old.
Mario is **5** years older.
How old is Mario?

8) Copy and complete in your jotter.
You may use counters to help you.

   a) 57 − 5 =

   b) 38 − 6 =

9) Copy and complete in your jotter:

   a) 31
       + 54

   b) 78
       − 42

10) Has this rectangle been cut in half?

   In your jotter, write **yes** or **no**.

11) Finley and Billie share 16 sweets.

Finley's sweets   Billie's sweets

a) Do they each get half the sweets?

b) How many sweets is half the sweets?

12) Is each shape cut into quarters?

In your jotter, write **yes** or **no**.

a)   b)

13) A pizza is cut into four equal pieces.
Kobe eats one of the pieces.
What fraction of the pizza does he eat?

14) Amir has 8 two pence coins.
He gives **4** of the coins to
Monica.
What **fraction** of the coins does
Amir have now?

15) How many equal ?

16) How many equal ?

17) How many equal ?

18) How many can I get for all these coins?

19) How many can I get for all these coins?

20) In your jotter, write the coins you might use to pay for each item **exactly**.

a) 45p

b) 69p

21) a) What is the day **after** Wednesday?

  b) What is the day **before** Sunday?

  c) What day is it 3 days **after** Friday?

  d) What day is it 7 days **after** Monday?

22) What season is **missing**?

Summer, winter, autumn, _____

23) a) What is the month **after** September?

  b) What month is it 3 months **after** March?

  c) What month is it 2 months **before** January?

24) a) What is the **7th** month of the year?

  b) What is the **3rd** month of the year?

25) In your jotter, write the time shown on each clock in words.

a)

b)

c)

26) I begin my bus journey to Parkside at 10 o'clock.
The journey takes **2 hours**.
At what time will I arrive at Parkside?

27) Use your ruler to measure each line, then write the lengths in your jotter.

a) ────────────

b) ──────────────

c) ───────────────

28) Use your ruler to draw and label lines that measure:

a) 7 cm   b) 5 cm   c) 13 cm.

29) Use your ruler to draw these shapes in your jotter:

a) a rectangle measuring 7 cm long and 5 cm wide

b) a square with sides measuring 5 cm.

30) Which fridge is **wider**?

 1    2

31) What would you use (a **ruler** or a **metre stick**) to measure the:

    a) width of your classroom   **b)** height of a book

    c) length of a garden   **d)** width of a laptop?

32) Which is **heavier**, a penguin or an elephant?

33) Which is **heavier**, the onion or the potato?

34) Put each list in order.
Start with the **heaviest**.

    a) 12 kg, 18 kg, 11 kg, 21 kg

    b) 10 kg, 17 kg, 21 kg, 99 kg

35) In your jotter, write the total mass, in kg, of the weights in each set.

    a)

    b)

36) Put the objects in order.

Start with the **lightest**.

37) In your jotter, draw and colour the next **2 shapes** in each pattern.

a)  ___ ___

b)  ___ ___

38) In your jotter, draw the next **2 shapes** in this pattern.

 __ __

39) To help you with these questions, write out all the letters of the alphabet in order, in your jotter.

Use your alphabet to help you continue each pattern.

a) P Q R S T U ___ ___

b) M L K J I H ___ ___

c) AC BD CE DF EG FH ___ ___

40) In your jotter, write the next **2 numbers** in each of the number patterns below.

    a)   22, 21, 20, 19, 18, ___, ___

    b)   90, 85, 80, 75, 70, ___, ___

    c)   12, 14, 16, 18, 20, ___, ___

    d)   30, 40, 50, 60, 70, ___, ___

41) In your jotter, write the numbers **1 more** and **1 less** than:

    a)   43   **b)**   57

    c)   90   **d)**   39.

42) What are the **missing** house numbers?

46   47   ◯   ◯   50   ◯

43) These cards are numbered **in order**.

47 is the card with the **smallest** number.

What number is on the **bottom** card?

44) What number does ⬚ stand for?

    a) $6 + \boxed{} = 9$ **b)** $5 + \boxed{} = 13$ **c)** $7 + \boxed{} = 12$

    d) $16 + \boxed{} = 17$ **e)** $9 - \boxed{} = 3$ **f)** $8 - \boxed{} = 5$

    g) $16 - \boxed{} = 9$ **h)** $20 - \boxed{} = 17$

45) What number does ◯ stand for?

a) $3 + ◯ = 11$ b) $◯ + 7 = 7$ c) $16 - ◯ = 7$

d) $15 + ◯ = 17$ e) $18 - ◯ = 14$ f) $◯ - 6 = 6$

g) $19 - ◯ = 9$ h) $◯ - 12 = 3$ i) $19 - ◯ = 0$

46) Look at the shapes below.

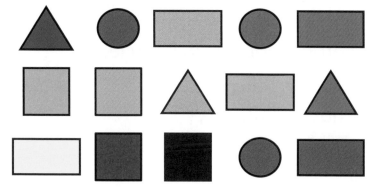

a) How many are triangles?

b) How many are circles?

c) How many are rectangles?

d) How many are squares?

47) In your jotter, draw a triangle like this.

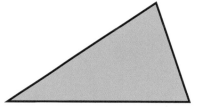

a) Write the word **side** next to each side.

b) Write the word **corner** next to each corner.

c) Mark each angle with an **arc**.

48) **a)** How many **sides** does this shape have?

    b) How many **angles** does it have?

    c) How many **corners** does it have?

    d) What is this shape called?

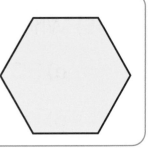

49) Use a ruler, when needed, to draw a shape with:

    a) 4 sides   **b)** 1 curved side.

50) **a)** How many sides does a square have?

    b) How many corners does a circle have?

51) Do these shapes have symmetry?
Use a mirror to check.

In your jotter, write **yes** or **no**.

a)    **b)**    **c)**

52) How many ways can each shape be folded to show symmetry?

a)    **b)**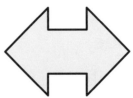

53) Which shape (A, B, C or D) should be added to the first shape to make it symmetrical?

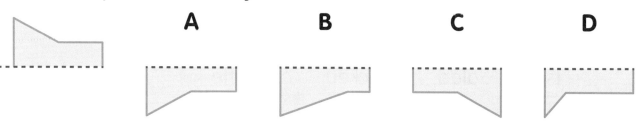

A   B   C   D

54) This table shows the number of pupils from a Primary 3 class who were absent from school during one week.

Day	Number of pupils absent
Monday	4
Tuesday	2
Wednesday	1
Thursday	3
Friday	5

a) How many pupils were absent on Friday?

b) How many pupils were absent on Tuesday?

c) How many **more** pupils were absent on Monday than on Tuesday?

d) How many **fewer** pupils were absent on Wednesday than on Friday?

e) How many pupils were absent **altogether** that week?

55) Two friends visited a restaurant.
The table shows what they ate.

	Starter	Main course	Dessert
Isaac	Salad	Chicken	Trifle
Oliver	Salmon	Omelette	Cheesecake

a) Who had **salmon** for their starter?

b) What did Isaac have for his main course?

c) What did each of them have for dessert?

56) In your jotter, copy the Carroll diagram below.

	Food	Not food
Red		
Not red		

Draw each object in the correct place in the diagram.

A   B   C   D   E

57) In your jotter, write down the shape the arrow will point to if it moves:

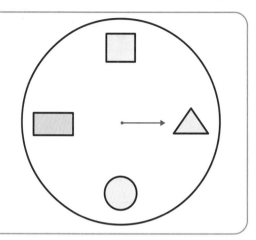

a)   a half turn clockwise

b)   a quarter turn anticlockwise

c)   a quarter turn clockwise.

58) In your jotter, write down the grid reference of:

a)   the square

b)   the triangle

c)   the circle.

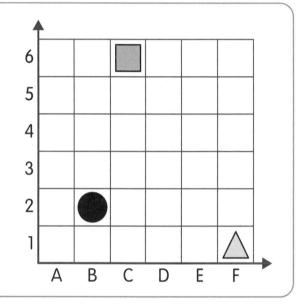

59) Milo follows a path over some paving slabs. In your jotter, write directions for the journey that Milo makes. Use **forward, turn right, turn left**.

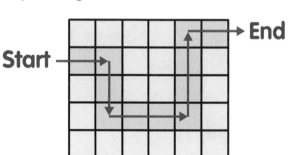

Whole numbers 1
Whole numbers to 1000

Numbers bigger than 100

💡 I will learn to understand place value for numbers from 100 to 1000.

125 counters can be arranged like this.

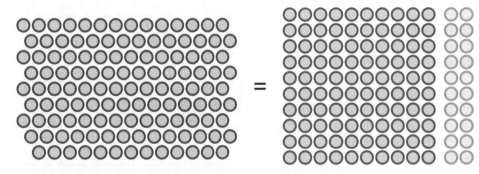

hundreds	tens	ones
1	2	5

		hundreds	tens	ones		
125	means	**1**	**2**	**5**	=	1 hundred, 2 tens and 5 ones.

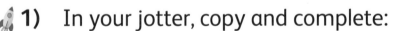
Exercise 1

🚀 1) In your jotter, copy and complete:

a) 237 = 2 hundreds, ___ tens and ___ ones

b) 418 = ___ hundreds, 1 ten and ___ ones.

2) In your jotter, write these numbers in hundreds, tens and ones:

a) 657 = ___ hundreds, ___ tens and ___ ones

b) 902 = ___ hundreds, ___ tens and ___ ones

c) 760 = ___ hundreds, ___ tens and ___ ones.

3) In your jotter, copy and complete with hundreds, tens or ones:

a) 5 = 5 ___

b) 50 = 5 ___ and 0 ___

c) 504 = 5 ___, 0 ___ and 4 ___

d) 540 = 5 ___, 4 ___ and 0 ___

4) In your jotter, write these numbers in hundreds, tens and ones:

a) 300   b) 602   c) 275

5) In your jotter, copy and complete:

a) 823 = 800 + 20 + ___   b) 238 = 200 + ___ + ___

c) 328 = ___ + ___ + 8   d) 302 = ___ + ___

⭐ 6) What do these digits stand for in the number 597? In your jotter, write **hundreds**, **tens** or **ones**.

a) 9   b) 7   c) 5

After 99 is:

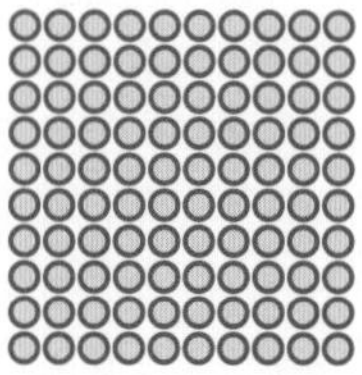

= 1 hundred = 100

Then 101, 102, 103, 104, 105 …

After 199 is:

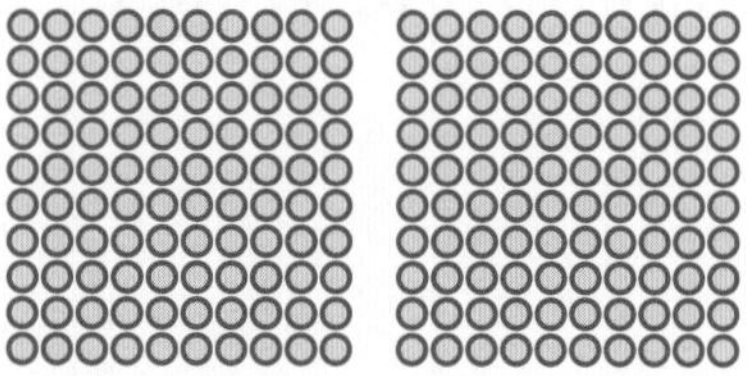

= 2 hundreds = 200

Then 201, 202, 203, 204, 205 …

After 299 is:

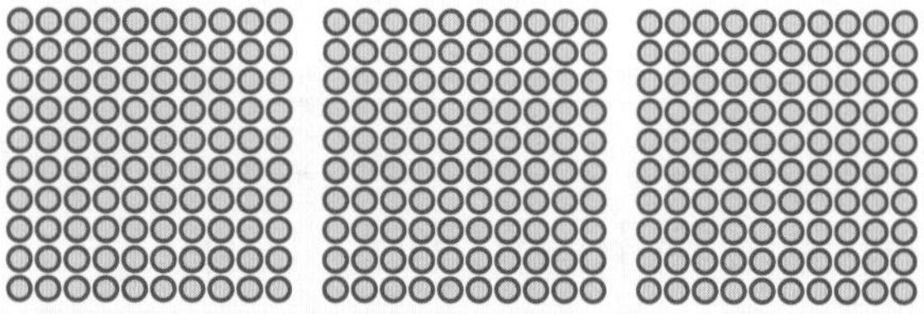

= 3 hundreds = 300

Then 301, 302, 303, 304, 305 …

And so on …

After 899 is:

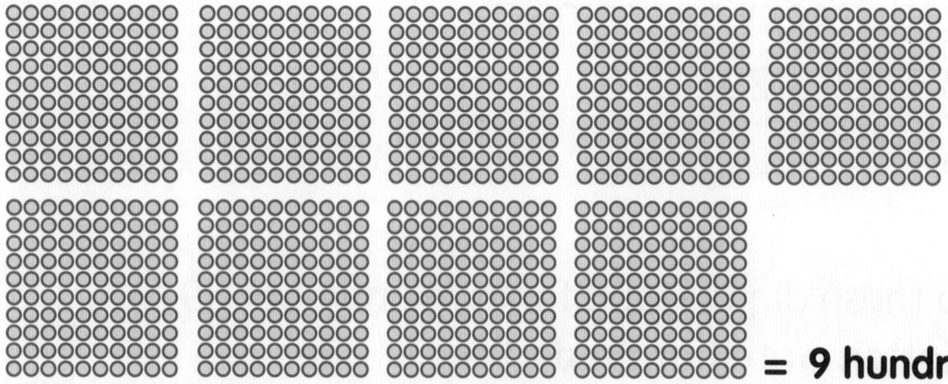

= 9 hundreds = 900

Then 901, 902, 903, 904, 905 …

The number after 999 is called **one thousand**.

It is written as **1000**.

10 hundreds = 1000

Exercise 2

1) In your jotter, write all the **missing** numbers.

   a)   168, ___, 170, ___, ___, 173

   b)   417, ___, 419, ___, ___, ___

   c)   995, ___, ___, 998, ___, ___

2) In your jotter, write all the **missing** numbers.

   a)   370, 369, ___, 367, ___, ___

   b)   221, ___, ___, 218, ___, ___

   c)   1000, ___, ___, 997, ___, ___

3) What is 1 more than:

   a)   410   b)   166

   c)   789   d)   999?

4) What is 1 less than:

   a)   628   b)   562

   c)   931   d)   870?

5) You can count in **ones** on a number line.
What number is the arrow pointing to?

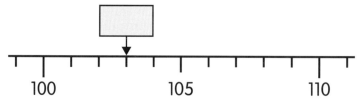

100   105   110

6) You can count in **tens** on a number line.
What number is each arrow pointing to?

a)

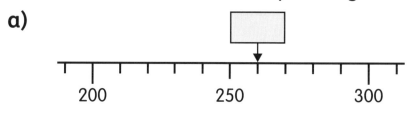

200   250   300

b)

700   800

7) You can count in **hundreds** on a number line.
What number is the arrow pointing to?

0   500   1000

⭐ **8)** Jo pays for a magazine.
She uses the exact amount:
four £1 coins, three 10p coins and nine 1p coins.
How much is the magazine?
Write your answer as ____ p.

Reading and writing numbers in words

 I will learn to read and write numbers in words.

In this section, you will learn about using words for numbers.

Digits are the numbers 0, 1, 2, 3, 4, 5, 6, 7, 8, 9.

Seven hundred and eighty-nine in **digits** is 789.

354 in **words** is three hundred and fifty-four.

Exercise 3

 1) In your jotter, copy the number from the box that is:

a) sixty-three

b) one hundred and thirty-six

c) one hundred and sixty-three

d) three hundred and thirty-six

e) one hundred and three.

> 136   63
>
> 336
>
> 163   103

2) In your jotter, write these numbers in **digits:**

a) twenty-three   b) thirty-eight   c) forty-one

d) seventy-five   e) sixty-two   f) ninety

g) twelve   h) eighty   i) fifty-nine

3) In your jotter, write these numbers in **digits**:

a) two hundred and forty-three

b) seven hundred and fourteen

c) nine hundred and seventy

d) eight hundred and eight

4) Copy and complete these numbers in words:

a) 57 is _____-seven

b) four hundred and _____ is 413

c) _____ and twenty is 920

d) 601 is _____ and _____

e) _____ and _____-nine is 739.

5) In your jotter, write these numbers in words:

a) 58   b) 46   c) 19   d) 70

e) 88   f) 311   g) 504   h) 1000

Let's try this!

Matching pairs

On cards, write the numbers 2, 14, 56, 137, 280 and 325.

Then, on different cards, write these numbers in words.

A game for 2 players.

1) Mix up the number and word cards and place them face down so you cannot see what is on them.

2) Take it in turns to turn over any 2 cards.

   If the 2 cards show the same number in **words** and in **digits**, pick them up.

   If the 2 cards do not show the same number, then turn them face down again.

   When all twelve cards are gone, the person the most cards is the winner.

Make the game harder by including cards with these six numbers written in **words** and **digits**: 36, 62, 141, 278, 305, 1000.

Revisit, review, revise

1) In your jotter, copy and complete:

   a) 512 = ____ hundreds, ____ tens and ____ ones

   b) 388 = 3 _____, 8 ____ and ____ ones

   c) 646 = 600 + ____ + 6

   d) 901 = 90 + ____

2) What does the 2 stand for in the number 729?
   In your jotter, write **hundreds**, **tens** or **ones**.

3) a) What number is 1 more than 88?

    b) What number is 1 less than 38?

4) In your jotter, write all the **missing** numbers.

    a) 462, 461, ____, ____, 458

    b) 128, ____, ____, 131, ____

5) What number is each arrow pointing to?

    Write the answers in your jotter.

    a)

    b)

    c)

6) a) In your jotter, write these numbers in **digits**:

        i) sixty-three   ii) one hundred and sixty-two

    b) In your jotter, write these numbers in **words**:

        i) 78   ii) 409   iii) 753   iv) 1000

② Symmetry
More symmetry

Symmetry in the real world

 I will learn to identify symmetry in the real world.

Symmetrical shapes can be found in the **real world**.

This shell, butterfly and flower show how symmetry can be found in nature.

Man-made things around us, like road signs, can also show symmetry.

Exercise 1

 1) Use a **mirror** to check that the pictures in the box above are symmetrical.

2) Do these pictures have symmetry?
In your jotter, write **yes** or **no**.
Use a mirror to help you decide.

a)

b)

c)

d)

e)

f)

g)

h)

i)

j)

k)

l)

3) Are these pictures symmetrical?
   In your jotter, write **yes** or **no**.

a)

b)

c)

d)

e) **30**

f)

4) In your jotter, write a list of things that have symmetry in:

   a) the classroom

   b) your house.

5) Collect some pictures of things that are symmetrical and make a **poster** for your classroom wall.

Revisit, review, revise

1) In your jotter, write five things in the real world that have symmetry.

2) Draw a picture of any road sign that does **not** have symmetry.

3 Whole numbers 2
More adding

Adding hundreds

💡 I will learn to add whole numbers with 3 digits (no 'carrying').

423 + 162 =

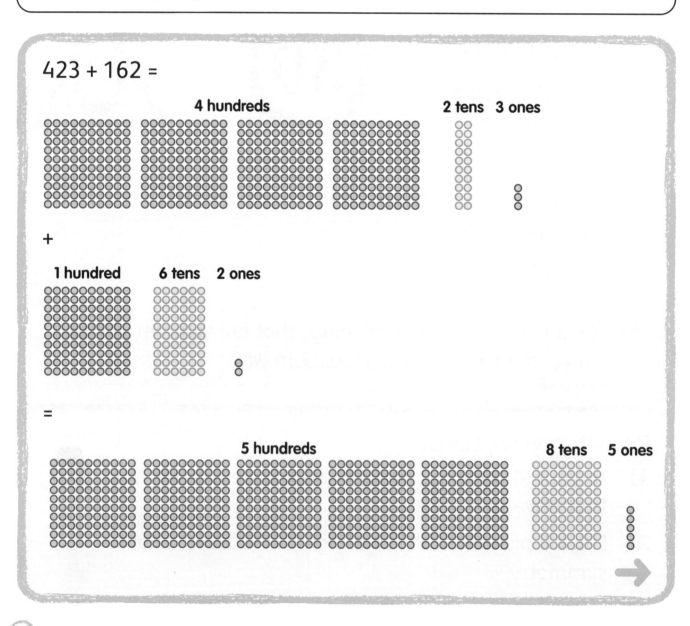

You can show this as:

hundreds	tens	ones
4	2	3

+

hundreds	tens	ones
1	6	2

=

hundreds	tens	ones
5	8	5

```
hundreds   tens   ones
   4        2       3
+  1        6       2
_____
   5        8       5
_____
```

or

```
 h  t  o
 4  2  3
+1  6  2
_____
 5  8  5
```

> When adding, line up the hundreds, line up the tens and line up the ones under each other.

```
400 + 100 = 500 ⎫
 20 +  60 =  80 ⎬ 585
  3 +   2 =   5 ⎭
```

You can also use number lines to add:

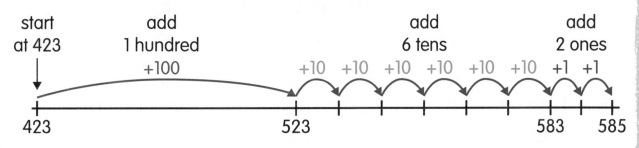

start at 423   add 1 hundred   add 6 tens   add 2 ones

+100   +10 +10 +10 +10 +10 +10   +1 +1

423   523   583   585

Exercise 1

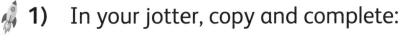

1) In your jotter, copy and complete:

a) $614 = 600 + 10 +$ ___

b) $285 =$ ___ $+$ ___ $+ 5$

c)

```
    hundreds    tens    ones
        6         1
  +    ___       ___      5
  _____
       ___       ___      9
```

2) Line up the hundreds, tens and ones under each other, then work out these additions:

a) 104 + 123   b) 429 + 210   c) 315 + 134

d) 506 + 391   e) 572 + 225   f) 609 + 140

g) 148 + 531   h) 482 + 311   i) 501 + 198

j) 345 + 604   k) 714 + 273   l) 273 + 714

3) Copy and complete to work out 600 + 300:
6 hundreds + 3 hundreds = ____ hundreds = ____00

4) Mentally (without writing anything), work out:

a) 100 + 400   b) 700 + 200   c) 400 + 12

d) 73 + 800   e) 200 + 160   f) 317 + 600

5) Line up the hundreds, tens and ones under each other, then work out these additions:

a) 637 + 52   b) 946 + 12   c) 814 + 72

d) 25 + 413   e) 71 + 128   f) 8 + 231

6) Copy and complete to work out 420 + 30:
4 hundreds + 2 tens + 3 tens = ____ hundreds + ____ tens
= ____0

7) Mentally (without writing anything), work out:

a) 600 + 30   b) 510 + 10   c) 260 + 30

d) 373 + 10   e) 415 + 60   f) 70 + 310

8) A laptop costs £380.
A printer costs £200.
What is the total cost of the laptop
and printer?

9) There are 80 pages in Seb's book and 612 pages in his
Mum's book.
How many pages in total in Seb's book and his Mum's book?

10) 508 seagulls are on a cliff.
191 puffins are there too.
How many birds in total on the cliff?

11) A play is on for 3 nights.
210 people watch on the first night.
73 people watch on the second night.

a) How many people watched the play on the first and
second night?

b) 216 people watch on the third night.
A newspaper says: 'More than 500 people watched
the play'.
Is the newspaper correct?
In your jotter, write **yes** or **no**.

Adding with carrying

 I will learn to add 1-digit and 2-digit whole numbers with 'carrying'.

25 + 9 =

You can use number lines to add:

start
at 25

add
9 ones

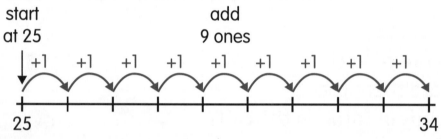

25   34

Looking to the **next 10 numbers** can help when using number lines.

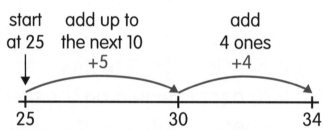

start
at 25

add up to
the next 10
+5

add
4 ones
+4

25   30   34

You can use ten frames to add:

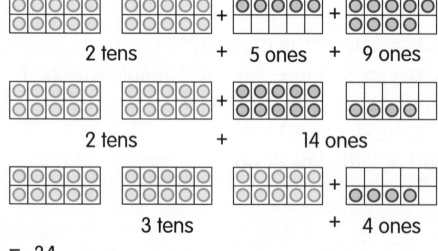

2 tens   +   5 ones   +   9 ones

2 tens   +   14 ones

3 tens   +   4 ones

= 34

You can show this as:

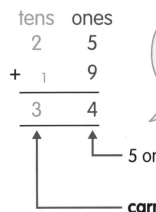

```
tens   ones
  2      5
+  1     9
  3      4
```

Remember, when adding, line up the tens and line up the ones.

5 ones + 9 ones = 14 ones
= 1 ten + 4 ones

carry 1 ten into the tens column.

Example

Work out 58 + 7.
Don't forget to add the number you have carried!

```
tens   ones
  5      8
+  1     7
  6      5
```

Line up the 7 below the 8.

Exercise 2

 1) Copy and complete these additions.
Use number lines or ten frames to help you.

a) 49 + 3	**b)** 27 + 4	**c)** 68 + 6
d) 77 + 7	**e)** 86 + 8	**f)** 39 + 9
g) 8 + 37	**h)** 9 + 37	**i)** 8 + 29

2) Line up the tens and the ones under each other, then work out these additions:

a) 38 + 8   **b)** 22 + 9   **c)** 79 + 4   **d)** 87 + 5

e) 88 + 7   **f)** 31 + 9   **g)** 6 + 59   **h)** 9 + 77

3) Summer buys a packet of fruit gums and a chew.
How much does she spend?

28p   6p

4) Sam has 46p in his money box.
Alice has 9p in hers.
How much do they have altogether?

5) A café sells 67 glasses of apple juice and 4 glasses of orange juice.
How many glasses of juice is that altogether?

6) In a zoo, there are 39 lions and 9 tigers.
How many lions and tigers altogether?

7) A bus can carry 58 people.
A taxi can carry 5 people.
How many people in total can be carried in a full bus and a full taxi?

⭐ **8)** Here are the prices of tickets to a show:

Adult	£18
Senior	£7
Child	£5

What is the total cost of 1 adult, 1 senior and 1 child ticket?

Let's try this!

Add 8 and carry

This is a game for two players.
You will need two dice.

Take it in turns.

1) Throw the dice.

Use your throw to make two 2-digit numbers.

Write them in your jotter.

For this throw, you would write down 15 and 51.

2) Add 8 to each of your numbers.

For this throw you would work out 15 + 8 and 51 + 8.

Scoring

Score 1 point for any answer greater than 50.

Lose 1 point if you don't need to carry into the tens column in your addition.

Who has the bigger score after five turns each?

Revisit, review, revise

1) In your jotter, work out these additions:

   a) 57 + 4   b) 18 + 7

   c) 9 + 64   d) 78 + 6

2) In your jotter, work out these additions:

   a) 375 + 523   b) 184 + 615

   c) 206 + 191   d) 501 + 408

   e) 713 + 41   f) 62 + 117

3) Mentally (without writing anything) work out:

   a) 200 + 500   b) 471 + 100

   c) 630 + 20   d) 923 + 50

4) Ranjet has 106 planes.
   Flyby has 232 planes.
   How many planes **altogether**?

5) Lucas spends £28 on Monday and £7 on Tuesday.
   How much does he spend by the end of Tuesday?

6) Frankie travels 147 miles from Glasgow to Aberdeen,
   then 202 miles to Wick.
   How far does he travel **in total**?

7) Sara has 785 bricks.
   She needs another 112 bricks to finish her job.
   How many bricks does she need altogether for the job?

4 Money
Money and decimals

Writing money using decimals

💡 **I will learn to handle and use decimal money up to £1.**

£1 can be written as £1.00.

The arrow is pointing to a **decimal point**.

93p can be written as £0.93.

52p can be written as £0.52.

30p can be written as £0.30.

7p can be written as £0.07.

When working with money, always have two numbers to the right of the decimal point.

Exercise 1

 1) In your jotter, write these amounts using a decimal point. (For example: 37p = £0.37)

a) 95p	**b)** 36p	**c)** 20p
d) 13p	**e)** 99p	**f)** 100p
g) 3p	**h)** 9p	**i)** 2p

2) In your jotter, write each of these as pence without a decimal point:

a) £0.45   **b)** £0.72   **c)** £0.83

d) £0.21   **e)** £0.50   **f)** £0.75

g) £1.00   **h)** £0.04   **i)** £0.05

Ninety-four pence can be written as 94p or £0.94.

---

3) In your jotter, write each amount in two ways (as above).

a) seventy-one pence   **b)** twenty-two pence

c) sixty pence   **d)** six pence

Money up to £5

 I will learn to use and handle money up to £5.

This is a £2 coin:

Jamie has one of each coin:

£2   £1   50p   20p   10p   5p   2p   1p

To work out how much Jamie has, add together the pounds first:

£2 + £1 = £3

Then add together the pence:

50p + 20p + 10p + 5p + 2p + 1p = 88p

Altogether Jamie has £3.88.

Exercise 2

1) How much does each person have?

a) Katie

b) Jackson

c) Otis

d) Lyla

£1.00 is the same as 100p.

£2.00 is the same as 200p.

£2.54 can be written as 254p.

£3.42 can be written as 342p.

Jen has these coins:

| £2 | £1 | 50p | 20p | 10p | 2p |

To work out how much Jen has, add together the pounds and pence separately:

£2 + £1 = £3

50p + 20p + 10p + 2p = 82p

82p can be written as £0.82.

Jen has £3 + £0.82 = £3.82.

2) In your jotter, write these amounts using a decimal point. (For example: 215p = £2.15)

    a) 194p   **b)** 230p   **c)** 401p

    d) 499p   **e)** 317p   **f)** 109p

3) In your jotter, write these amounts as pence. (For example: £3.20 = 320p)

    a) £4.12   **b)** £2.10   **c)** £3.01

    d) £4.50   **e)** £3   **f)** £0.99

4) How much does each person have?

a) Zain

b) Val

c) Shay

d) Jasmine

e) Kwame

5) Who has more money?
Len has

Fran has

6) Draw the coins you could use to make £4.31.

Let's try this!

How many different ways can you find of making £5 using coins?

Use as many coins as you like.

Try to draw at least 5 different ways.

Adding and subtracting money up to £5

Alin has £5.00.

He buys this snack and drink.

How much money does he have left?

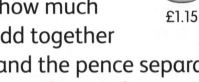

£1.80

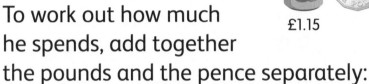

£1.15

To work out how much he spends, add together the pounds and the pence separately:

£1 + £1 = £2

50p + 20p + 10p + 10p + 5p = 95p

He spends £2.95.

Use a number line to work out how much money he has left:

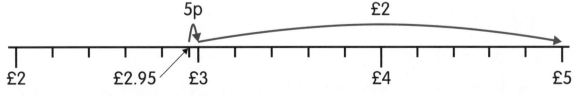

Alin has £2.05 left.

Exercise 3

 1) Here is how much money Jon and Mae have.

Jon

Mai

    a) How much does Jon have?

    b) How much does Mae have?

    c) How much do they have altogether?

2) How much does it cost to buy:

    a) an apple and an orange

    b) water and an apple

    c) an apple, an orange and water

        30p   50p   £1.10

    d) Bill has £4.
        He buys an apple, an orange and water.
        How much does he have left?

3) Here is how much money Peter and Kai have:

Peter

Kai

a) How much do Peter and Kai have altogether?

b) How much **more** does Peter have than Kai?

4) Layla has £3.40.

Her gran gives her 47p.

How much does she have now?

5) Zak buys a drink for £2.45 and a biscuit for £1.53.

How much does he spend?

Revisit, review, revise

1) In your jotter, write these amounts using a **decimal point**.
(For example: 54p = £0.54)

    a) 62p   **b)** 9p   **c)** fifty-one pence

    d) ninety pence   **e)** 210p   **f)** 309p

2) How much money does each person have?

    a) Maya

b) Thomas

c) Sadie

3) In your jotter, write these amounts using a **decimal point**. (For example: 54p = £0.54)

a) 257p

b) six pounds and one penny

4) Amy buys a book for £2.20 and a drink for 75p.
How much does she spend altogether?

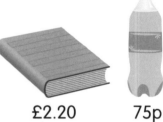

£2.20   75p

5) Jon has £5.
He buys pencils.
How much does he have left?

£2.73

Whole numbers 3
More subtracting

Subtracting hundreds

> 💡 I will learn to subtract from a 3-digit number (no 'exchanging' or 'borrowing').

465 – 123 =

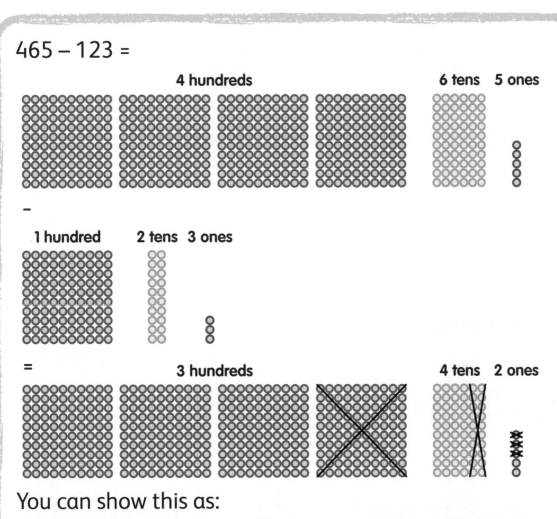

You can show this as:

hundreds	tens	ones		hundreds	tens	ones		hundreds	tens	ones
4	6	5	–	1	2	3	=	3	4	2

hundreds	tens	ones
4	6	5
− 1	2	3
3	4	2

or

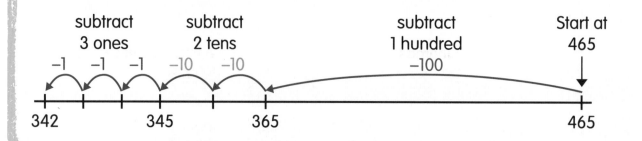

When subtracting, line up the hundreds, line up the tens and line up the ones under each other.

```
h  t  o
4  6  5
-1  2  3
3  4  2
```

$400 - 100 = 300$
$60 - 20 = 40$ } 342
$5 - 3 = 2$

You can also use a number line to subtract:

subtract 3 ones   subtract 2 tens   subtract 1 hundred   Start at 465

−1 −1 −1   −10 −10   −100

342   345   365   465

Exercise 1

 1) In your jotter, copy and complete:

a) $678 = 600 + 70 +$ ____

b) $237 =$ ____ $+$ ____ $+ 7$

c)

hundreds	tens	ones
6	7	___
− ___	___	7
___	___	1

2) Line up the hundreds, tens and ones under each other, then work out these subtractions:

    a) 375 – 212   **b)** 398 – 194

    c) 406 – 102   **d)** 499 – 357

    e) 575 – 263   **f)** 695 – 344

    g) 728 – 413   **h)** 797 – 525

    i) 746 – 214   **j)** 999 – 666

    k) 622 – 601   **l)** 254 – 253

3) In your jotter, copy and complete to work out 800 – 500:

8 hundreds – 5 hundreds = __ hundreds

= __00

4) Mentally (without writing anything) work out:

    a) 600 – 100   **b)** 500 – 200   **c)** 900 – 700

    d) 812 – 100   **e)** 390 – 190   **f)** 501 – 101

5) Line up the hundreds, tens and ones under each other, then work out these subtractions:

    a) 193 – 82   **b)** 768 – 17   **c)** 592 – 61

    d) 634 – 23   **e)** 827 – 13   **f)** 286 – 4

6) In your jotter, copy and complete to work out 380 – 60:

3 hundreds + 8 tens – 6 tens = __ hundreds + __ tens

= ____0

7) Mentally (without writing anything) work out:

  a) 540 – 40   **b)** 270 – 30   **c)** 950 – 40

  d) 880 – 50   **e)** 390 – 70   **f)** 675 – 75

8) There are 382 tadpoles in a pond.

211 tadpoles turn into frogs.

How many tadpoles have not yet turned into frogs?

9) 350 people are asked if they like curry.
40 say they do not like curry.
How many do like curry?

10) Ed pays £999 for a television.
His fridge cost £720 less than that.

  a) How much was his fridge?

  b) His vacuum cleaner cost £175 less than the fridge.
    How much was his vacuum cleaner?

Subtracting with exchanging

 I will learn to subtract a 1-digit number from a 2-digit number with 'exchanging'.

34 – 8 =

You can use number lines to subtract:

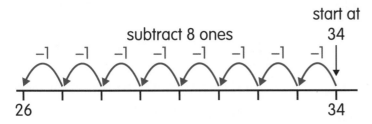

Looking to the **last 10 numbers** can help when using number lines:

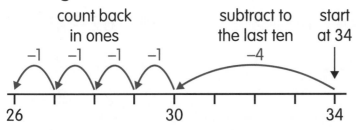

You can use ten frames to subtract:

34 – 8 =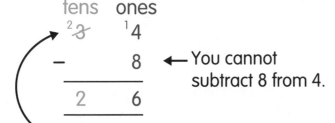

2 tens   6 ones

= 26

You can show this as:

tens   ones

²3̶   ¹4

–   8   ← You cannot subtract 8 from 4.

2   6

Remember, when subtracting, line up the tens and line up the ones.

exchange 1 ten for 10 ones and carry them into the ones column, now there are 2 tens left and 14 ones – 8 ones

Exercise 2

 1) Copy and complete these subtractions.
Use number lines or ten frames to help you.

a) 22 – 4 **b)** 31 – 5 **c)** 27 – 8

2) In your jotter, work out these subtractions:

a) $^3\cancel{4}\ ^12$
   –   3
  ‾‾‾‾‾

b) $^2\cancel{3}\ ^11$
   –   8
  ‾‾‾‾‾

c) $^5\cancel{6}\ ^16$
   –   7
  ‾‾‾‾‾

3) Line up the tens and the ones under each other,
then work out these subtractions:

a) 44 – 5 **b)** 80 – 2 **c)** 62 – 4

d) 35 – 7 **e)** 71 – 3 **f)** 53 – 9

g) 76 – 8 **h)** 91 – 6 **i)** 27 – 8

4) There are 25 cakes on a plate.
8 of them are eaten.
How many are left?

5) There are 67 people at the station.
9 are waiting for the Ayr train.
How many are not waiting for the Ayr train?

6) Irene has a book of 93 raffle tickets.
She sells 9 tickets to her friends.
She sells 4 tickets to her family.
How many tickets does she have left?

Let's try this!

Dice exchange

This is a game for two players.
You will need three dice.

Take it in turns.

1) Throw two dice.

   Use your throw to make two 2-digit numbers. Write them in your jotter.

   For this throw, you would write down 52 and 25.

2) Throw the third dice.

   Subtract the number on this throw from each of your 2-digit numbers.

   For this throw you would work out 52 – 4 and 25 – 4.

Scoring

Score 3 points for any answer that has a digit 9.

Score 2 points for any answer that has a digit 8.

Score 1 point for any answer that has a digit 7.

For this throw, you would score 2 points because
52 – 4 = 48, so the answer has a digit 8.

Who has the bigger score after five turns each?

Revisit, review, revise

1) In your jotter, copy and complete these subtractions:

   a) $\begin{array}{r} 62 \\ -\ 8 \\ \hline \end{array}$ **b)** $\begin{array}{r} 41 \\ -\ 5 \\ \hline \end{array}$ **c)** $\begin{array}{r} 84 \\ -\ 5 \\ \hline \end{array}$

2) In your jotter, work out these subtractions:

   a) $74 - 6$ **b)** $63 - 7$

   c) $31 - 8$ **d)** $26 - 9$

3) In your jotter, work out these subtractions:

   a) $385 - 152$ **b)** $846 - 504$ **c)** $692 - 31$

4) Mentally (without writing anything) work out:

   a) $800 - 500$ **b)** $629 - 100$

   c) $307 - 107$ **d)** $760 - 60$

   e) $450 - 30$ **f)** $238 - 38$

5) There are 175 shops in a shopping centre.
61 sell food.
How many do not sell food?

6) Sally saves £578.

She spends £106.

How much is left?

6 Whole numbers 4
Rounding and estimating

Rounding to the nearest 10 and 100

💡 I will learn to round a whole number to the nearest 10 and 100.

Here is a number line:

The arrow points to 48.

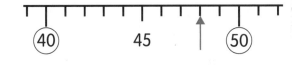

The **last ten** before 48 is 40. The **next ten** after 48 is 50.

48 lies between 40 and 50.

48 is **closer** to 50 than 40.

We say that 48, **rounded to the nearest 10**, is 50.

Exercise 1

1) Look at the number line. In your jotter, write the **missing** numbers.

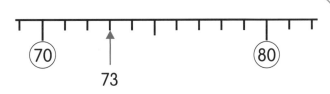

a) The **last ten** before 73 is ___

b) The **next ten** after 73 is ___

c) 73 lies between 70 and ___

d) 73 is closer to ___ than ___

e) 73 rounds to ___ (to the nearest 10).

2) In your jotter, copy and complete:

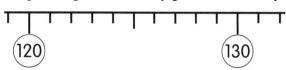

    a) 129 lies between 120 and ____

    b) 129 is closer to ____ than ____

    c) 129 rounds to ____ (to the nearest 10).

3) In your jotter, copy and complete:

    a) 24 lies between 20 and ____

    b) 24 is closer to ____ than ____

    c) 24 rounds to ____ (to the nearest 10).

4) Imagine these numbers on a number line and decide what each one rounds to (to the nearest 10). In your jotter, copy and complete:

    a) 87 lies between 80 and ____
        It is closer to ____

    b) 133 lies between ____ and 140
        It is closer to ____

    c) 458 lies between 450 and ____
        It is closer to ____

    d) 902 lies between 900 and ____
        It is closer to ____

When we round to the **last ten**, it is called **rounding down**.

When we round to the **next ten**, it is called **rounding up**.

35 lies between 30 and 40.

It is the same distance from 30 to 35 and 35 to 40.

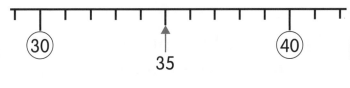

35 is not closer to 30 or 40.

When a number **ends in 5**, we **round up**.

We say that 35, **rounded to the nearest 10**, is 40.

---

5) Round these numbers to the **nearest 10**:

a) 37   b) 71   c) 18

d) 63   e) 159   f) 141

g) 316   h) 534   i) 45

Here is a number line:
The arrow points to 370.

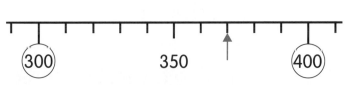

We say that 370, **rounded to the nearest 100**, is 400.

---

6) Look at the number line. In your jotter, write the **missing** numbers.

a) The **last hundred** before 620 is ____

b) The **next hundred** after 620 is ____

c) 620 lies between ____ and ____

d) 620 is closer to ____ than ____

e) 620 rounds to ____ (to the nearest 100).

7) Look at the number line in question 6.
Point to where you think 671 is.

a) Is 671 closer to 600 or 700?

b) Round 671 to the nearest 100.

8) Imagine these numbers on a number line and decide what each one rounds to (to the nearest 100).
In your jotter, copy and complete:

a) 280 lies between ____00 and ____00.
It is closer to ____

b) 840 lies between ____00 and ____00.
It is closer to ____

c) 312 lies between ____00 and ____00.
It is closer to ____

When we round to the **last hundred**, it is called **rounding down**.

When we round to the **next hundred**, it is called **rounding up**.

250 lies between 200 and 300.

It is the same distance from 200 to 250 and from 250 to 300.

250 is not closer to 250 or 300.

When a number **ends in 50**, we **round up**.

We say that 250, **rounded to the nearest 100**, is 300.

9) Round these numbers to the **nearest 100**:

    a) 120   **b)** 380   **c)** 690   **d)** 430

    e) 910   **f)** 560   **g)** 150   **h)** 750

10) a) The height of this pot is 33 centimetres. Round this to the nearest 10 cm.

    b) A builder earns £185 per day. Round this to the nearest £100.

    c) A shoe shop has 628 pairs of shoes. Round the number of pairs of shoes to the nearest 10.

Estimating by rounding

 I will learn to estimate answers by rounding numbers to the nearest 10.

You can round numbers in a calculation to the **nearest 10** to estimate the answer.

Example 1

256 + 32

is about 260 + 30 = 290.

Example 2

951 − 428

is about 950 − 430 = 520.

Exercise 2

 1) a) In your jotter, write 88 rounded to the nearest 10.

    b) In your jotter, write 42 rounded to the nearest 10.

    c) Use your answers to **a)** and **b)** to copy and complete:

       88 − 42 is about ____ − ____ = ____

2) In your jotter, copy and complete to estimate the answer to the nearest ten:

a) $56 + 38$

is about

$60 + 40 =$ ____

b) $22 + 77$

is about

$20 +$ ____ $=$ ____

c) $18 + 49$

is about

____ $+ 50 =$ ____

d) $31 + 45$

is about

____ $+$ ____ $=$ ____

e) $149 + 214$

is about

____ $+ 210 =$ ____

f) $727 + 146$

is about

____ $+$ ____ $=$ ____

3) In your jotter, copy and complete to estimate the answer to the nearest ten:

a) $73 - 24$

is about

$70 -$ ____ $=$ ____

b) $89 - 76$

is about

____ $-$ ____ $=$ ____

c) $152 - 27$

is about

$150 -$ ____ $=$ ____

d) $189 - 142$

is about

$190 -$ ____ $=$ ____

e) $454 - 199$

is about

____ $-$ ____ $=$ ____

f) $581 - 216$

is about

____ $-$ ____ $=$ ____

4) Mentally (without writing anything) estimate to the nearest ten:

a) $29 + 63$

b) $47 + 99$

c) $53 + 49$

d) $171 + 103$

e) $73 - 29$

f) $252 - 97$

⭐ **5)** A ticket to a show costs £119.

Estimate the total value of 2 tickets.

Let's try this!

You will need counters.
Work in pairs.

Calculation cloud

553 + 17

346 + 24   427 + 333

611 + 52   185 + 702   931 – 21

852 – 48

264 – 139

549 – 312

771 – 517

Estimated answer

890

370   910

570   760   240   120

500   800

660   250

1) Choose any calculation from the calculation cloud.

2) Estimate the answer and find it in the answer cloud.

3) Cover the calculation and its answer with counters.

4) One answer is left uncovered.
Which one?

Challenge

In your jotter, write a calculation for the uncovered estimated answer.

Revisit, review, revise

1) Look at this number line.

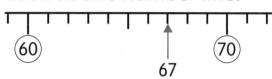

Round 67 to the nearest 10.

2) Round each number to the nearest 10.

a) 43 **b)** 69 **c)** 457 **d)** 285

3) Look at this number line.

Round 440 to the nearest 100.

4) Round each number to the nearest 100.

a) 810 **b)** 360 **c)** 750 **d)** 481

5) Estimate the answer to 248 + 149.
248 + 149 is about 250 + ____ = ____

6) Estimate the answer to each calculation.

a) 49 + 21 **b)** 239 + 424

7) Jason has 263 comics.
He sells 137 of them.
Estimate, to the nearest 10, how many comics he keeps.

8) A baker has 431 apples.
She uses 207 to make apple pies.
She uses the rest to make apple crumbles.
Estimate how many apples she uses for crumbles.

Shapes and objects
3D objects

Naming 3D objects

> 💡 I will learn to recognise and name some common 3D (solid) shapes.

In this section, you will learn about some common **3-dimensional** (**3D**) shapes.

You already know the names of some 2D shapes. These are the **flat shapes**: square, rectangle, circle and triangle.

square   rectangle   circle   triangle

In this chapter, you will meet the **solid shapes**: cube, cuboid, cone, cylinder and sphere.

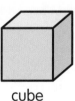 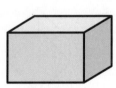

cube   cuboid   cone

cylinder   sphere

Here are three more 3D shapes:

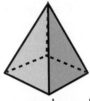

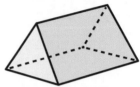

square-based pyramid   hemisphere   triangular prism

Exercise 1

1) In your jotter, write the names of these 3D shapes:

a)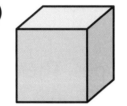

b)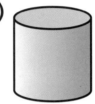

c)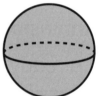

d)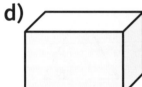

e)

2) In your jotter, write the names of these 3D shapes:

a)

b)

c)

d)

e)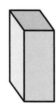

3) What are these 3D shapes called?

a)

b)

c)

4) Here are some 3D items found in the real world. In your jotter, write the name of each shape.

a)

b)

c)

d)

e)

f)

g)

h)

5) In your jotter, make a list of the 3D shapes used in each picture below.

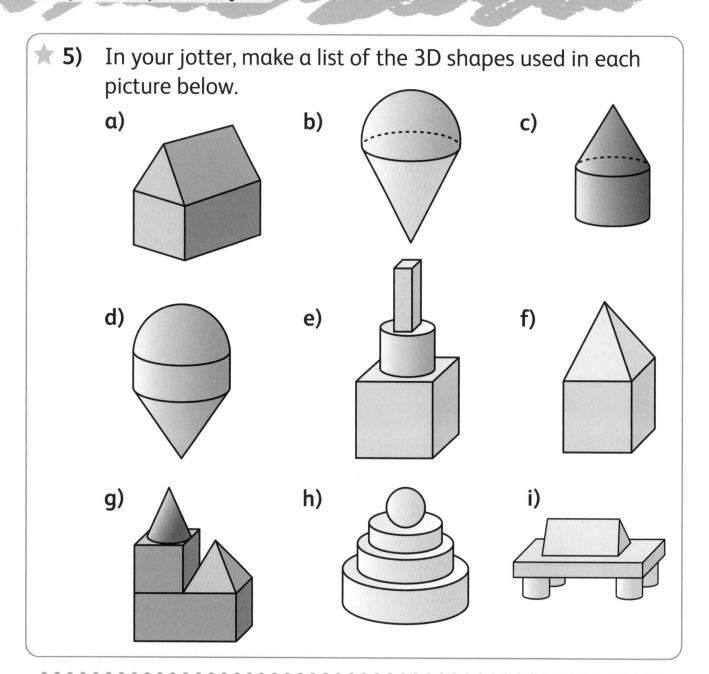

a)

b)

c)

d)

e)

f)

g)

h)

i)

Let's try this!

Go on a shape hunt in your classroom.

Collect all the 3D shapes you can find.

As a class, put your 3D shapes into groups of the same shape.

Making 3D objects

 I will learn to describe the faces of a 3-dimensional shape.

The flat surfaces of a 3-dimensional (3D) shape are called **faces**.

The faces of a 3D shape are 2-dimensional (2D) shapes.

Here is a cube:

The faces of a cube are all squares:

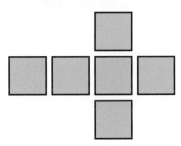

A cube has 6 square faces.

Some 3D shapes have curved faces.

Here is a cone:

One face is a circle. The other face is curved.

Exercise 2

 1) What shape are the faces of this cuboid?

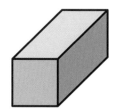

2) What shape are the faces of this cuboid?

3) In your jotter, make a list of the shapes of all the faces of these 3D shapes.

a)

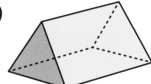

b)

4) Which of these shapes have a curved face?

a)

b)

c)

d)

e)

f)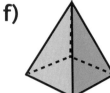

5) What 3D shape has these faces:

a) 6 squares

b) 6 rectangles

c) 2 triangles and 3 rectangles

d) 4 triangles and 1 square

e) 1 circle and 1 curved face

f) 1 curved face and 2 circles?

Faces, edges and vertices

💡 **I will learn to work out how many faces, edges and vertices a 3D shape has.**

Some 3D shapes have **edges** and **vertices** (corners).
Here is a cuboid:

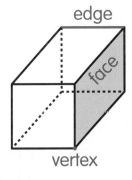

Count how many faces it has.
It has **6** faces.

An **edge** is where two faces meet.
Count how many edges this cuboid has.
Some are hiding round the back of the shape.
It has **12** edges.

A **vertex** (corner) is where edges meet.
Count how many vertices this cuboid has.
Some are hiding round the back of the shape.
It has **8** vertices.

Exercise 3

For this exercise, you will need a model of each shape:

- cube
- cone
- square-based pyramid
- cylinder.
- cuboid
- sphere
- triangular prism

1) Pick up your model cuboid.

a) Slide your finger along one of its faces and then the other faces.
How many faces does it have?

b) Press your finger against one of its vertices and then the others.
How many vertices does it have?

c) Run your finger along one of its edges and then the others.
How many edges does it have?

2) Look at your model cube.

a) How many faces does it have?

b) How many vertices does it have?

c) How many edges does it have?

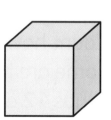

3) Look at your model square-based pyramid.

   a) How many vertices does it have?

   b) How many faces does it have?

   c) How many edges does it have?

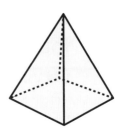

4) Look at your model triangular prism.

   a) How many edges does it have?

   b) How many vertices does it have?

   c) How many faces does it have?

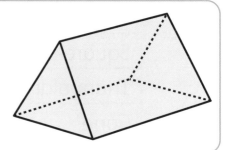

⭐ **5)** Look at your model cone.

   a) How many vertices does it have?

   b) How many edges does it have?

   c) How many faces does it have?

⭐ **6)** Look at your model sphere.

   a) How many faces does it have?

   b) How many vertices does it have?

   c) How many edges does it have?

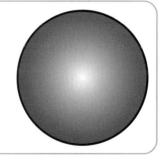

⭐ **7)** Look at your model cylinder.

   a) How many vertices does it have?

   b) How many edges does it have?

   c) How many faces does it have?

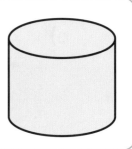

8) Copy and fill in this table to display all the results you have found in this exercise.

(Try not to look back at your answers.)

	Faces	Vertices	Edges
Cuboid	6		
Cube			
Square-based pyramid			
Triangular prism			
Cone			
Sphere			
Cylinder			

9) Three of your 3D shapes can be rolled.
Which three?

Revisit, review, revise

1) What are the names of these 3D shapes?

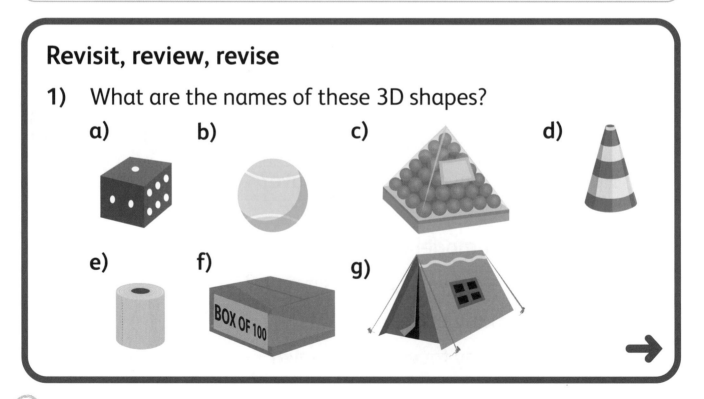

a)   b)   c)   d)

e)   f)   g)

BOX OF 100

2) List the 3D shapes shown in the picture and say how many of each shape you can see.

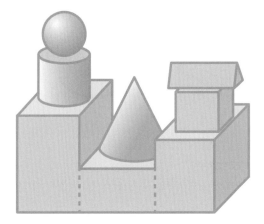

3) In your jotter, write down the shapes of the faces of a square-based pyramid.

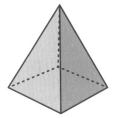

4) What 3D shape has these faces:

    a) 6 squares

    b) 2 triangles and 3 rectangles

    c) 6 rectangles

    d) 4 triangles and a square?

5) Here is a cuboid:

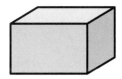

    a) How many faces does it have?

    b) How many vertices does it have?

    c) How many edges does it have?

6) Look at the pictures.

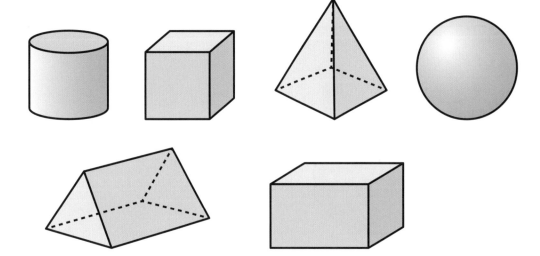

How many:

a) edges has a cube

b) vertices has a cuboid

c) edges has a square-based pyramid

d) vertices has a triangular prism?

8 Whole numbers 5
Multiplying by 2 and 3

Doubles and trebles

💡 **I will learn to double and treble numbers.**

In this array, there are 4 counters:

Now there are **two times** the number of counters:

We can say there is **double** the number of counters.

Count the counters.

Double 4 is 8.

In this ten frame, there are 7 ones:

→

Now there are **two times** the number of ones:

We can say there is **double** the number of ones.

We can rearrange this in ten and ones:

Double 7 is 14.

Exercise 1

1) In your jotter, copy this array two **times**.
What is **double** 3?

2) Copy this ten frame two **times**.
What is **double** 5?

3) Work out these **doubles**.
You can use counters or ten frames to help you.

a) **double** 2   b) **double** 10   c) **double** 6

4) In your jotter, write:

a) 11 rounded to the nearest 10

b) an estimate for **double** 11. (Use your answer to **a**.)

In this array, there are 4 counters:

Now there are **three times** the number of counters:

We can say there is **treble** the number of counters.

Count the counters:

Treble 4 is 12.

In this ten frame, there are 7 ones:

Now there are **three times** the number of ones:

We can say there is **treble** the number of ones.
We can rearrange this in tens and ones:

Treble 7 is 21.

Exercise 2

 1) In your jotter, copy this array three **times**. What is **treble** 2?

 2) Copy this ten frame three **times**. What is **treble** 5?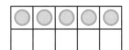

3) In your jotter, work out these **trebles**. You can use counters or ten frames to help you.

    a) treble 3   **b)** treble 10   **c)** treble 6

 **4)** **Double** 4. Then **treble** the answer.

2 times table

 I will learn the 2 times table and to multiply by 2.

2 lots of 1 counter = 2 counters.

We say 2 × 1 = 2.

2 **times** 1 is 2.

2 **multiplied by** 1 is 2.

2 lots of 2 counters = 4 counters.

We say 2 × 2 = 4.

2 **times** 2 is 4.

2 **multiplied by** 2 is 4.

2 lots of 3 counters = 6 counters.

We say $2 \times 3 = 6$.

2 **times** 3 is 6.

2 **multiplied by** 3 is 6.

2 lots of 4 counters = 8 counters.

We say $2 \times 4 = 8$.

2 **times** 4 is 8.

2 **multiplied by** 4 is 8.

The sign × is called the **times** or **multiplication** sign.

Exercise 3

1) Work out these **multiplications**.
   You may use counters to help you.

   a) 2 lots of 5 counters: $2 \times 5 =$ _____

   b) 2 lots of 6 counters: $2 \times 6 =$ _____

   c) 2 lots of 7 counters: $2 \times 7 =$ _____

   d) 2 lots of 8 counters: $2 \times 8 =$ _____

   e) 2 lots of 9 counters: $2 \times 9 =$ _____

   f) 2 lots of 10 counters: $2 \times 10 =$ _____

2) In your jotter, write the title '**2 times table**'.
Now copy and complete:

2 × 1 = 2   2 × 6 = ____

2 × 2 = ____   2 × 7 = ____

2 × 3 = ____   2 × 8 = ____

2 × 4 = ____   2 × 9 = ____

2 × 5 = ____   2 × 10 = ____

3) In your jotter, copy the grid.

1	2	3	4
5	6	7	8
9	10	11	12
13	14	15	16
17	18	19	20

Colour the numbers in the **2 times table**.

4) Jon is skip counting in 2s on this number line.

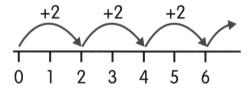

He counts 2, 4, 6.
In your jotter, write the next **7 numbers** that he says.

5) Try to learn the **2 times table** by heart.
Say to yourself, 2 × 1 = 2, 2 × 2 = 4, 2 × 3 = 6, 2 × 4 = 8, ...
Now say it aloud to a friend.

1 lot of 2 counters = 2 counters.

We say 1 × 2 = 2.

1 **times** 2 is 2.

1 **multiplied** by 2 is 2.

2 lots of 2 counters = 4 counters.

We say 2 × 2 = 4.

2 **times** 2 is 4.

2 **multiplied** by 2 is 4.

3 lots of 2 counters = 6 counters.

We say 3 × 2 = 6.

3 **times** 2 is 6.

3 **multiplied** by 2 is 6.

4 lots of 2 counters = 8 counters.

We say 4 × 2 = 8.

4 **times** 2 is 8.

4 **multiplied** by 2 is 8.

Notice that:

1 × 2 = 2   and   2 × 1 = 2

2 × 2 = 4   and   2 × 2 = 4

3 × 2 = 6   and   2 × 3 = 6

4 × 2 = 8   and   2 × 4 = 8.

These are both ways of writing the **2 times table**.

6) What numbers are **missing**?

a) ___ × 2 = 4   b) 2 × ___ = 8   c) ___ × 2 = 12

d) ___ × 2 = 18   e) 2 × ___ = 14   f) 2 × ___ = 16

g) 2 × ___ = 10   h) ___ × 2 = 20   i) 0 × 2 = ___

7) There are 4 nests.
Each nest has 2 chicks.
How many chicks altogether?

8) Every room in a house has 2 sockets.
The house has 7 rooms.
How many sockets in total?

★ **9)** Priya and Demi have 8 plums each.
How many plums do they have altogether?

When you **double** a number, you **multiply by 2**.

Count the counters.

Double 4 is 8.

4 × 2 = 8

⭐ **10)** In your jotter, copy and complete:

     a)   **double** 3 = 3 × 2 = ____

     b)   **double** 10 = 10 × 2 = ____

     c)   **double** 6 = ____ × ____ = ____

  ⭐ d)   **double** ____ = ____ × 2 = 10

  ⭐ e)   **double** ____ = ____ × ____ = 18

3 times table

 I will learn the 3 times table and to multiply by 3.

3 lots of 1 counter = 3 counters.

⊙ + ⊙ + ⊙ = ○ ○ ○

We say 3 × 1 = 3.

3 **times** 1 is 3.

3 **multiplied by** 1 is 3.

3 lots of 2 counters = 6 counters.

▯ + ▯ + ▯ = ○ ○ ○ / ○ ○ ○

We say 3 × 2 = 6.

3 **times** 2 is 6.

3 **multiplied by** 2 is 6.

3 lots of 3 counters = 9 counters.

▯ + ▯ + ▯ = ○ ○ ○ / ○ ○ ○ / ○ ○ ○

We say 3 × 3 = 9.

3 **times** 3 is 9.

3 **multiplied by** 3 is 9.

→

3 lots of 4 counters = 12 counters.

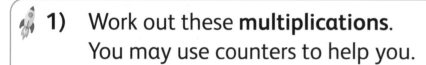

We say 3 × 4 = 12.

3 **times** 4 is 12.

3 **multiplied by** 4 is 12.

Remember, the sign × is called the **times** or **multiplication** sign.

Exercise 4

1) Work out these **multiplications**.
You may use counters to help you.

    a) 3 lots of 5 counters:
        3 × 5 = ____

    b) 3 lots of 6 counters:
        3 × 6 = ____

    c) 3 lots of 7 counters:
        3 × 7 = ____

    d) 3 lots of 8 counters:
        3 × 8 = ____

    e) 3 lots of 9 counters:
        3 × 9 = ____

    f) 3 lots of 10 counters:
        3 × 10 = ____

2) In your jotter, write the title '**3 times table**'.
Now copy and complete:

3 × 1 = ____   3 × 6 = ____

3 × 2 = ____   3 × 7 = ____

3 × 3 = ____   3 × 8 = ____

3 × 4 = ____   3 × 9 = ____

3 × 5 = ____   3 × 10 = ____

3) In your jotter, copy the grid.

1	2	3	4	5	6
7	8	9	10	11	12
13	14	15	16	17	18
19	20	21	22	23	24
25	26	27	28	29	30

Colour the numbers in the **3 times table**.

4) Jane is skip counting in 3s on this number line.
She counts 3, 6, 9.

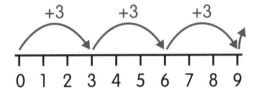

In your jotter, write the next **7 numbers** that she says.

5) Try to learn the **3 times table** by heart.
Say to yourself, $3 \times 1 = 3$, $3 \times 2 = 6$, $3 \times 3 = 9$, $3 \times 4 = 12$,
Now say it aloud to a friend.
(Remember to keep practising your **2 times table**.)

1 lot of 3 counters = 3 counters.

We say $1 \times 3 = 3$.
1 **times** 3 is 3.
1 **multiplied by** 3 is 3.

2 lots of 3 counters = 6 counters.

We say $2 \times 3 = 6$.

2 **times** 3 is 6.

2 **multiplied by** 3 is 6.

3 lots of 3 counters = 9 counters.

We say $3 \times 3 = 9$.

3 **times** 3 is 9.

3 **multiplied by** 3 is 9.

4 lots of 3 counters = 12 counters.

We say $4 \times 3 = 12$.

4 **times** 3 is 12.

4 **multiplied by** 3 is 12.

Notice that:

$1 \times 3 = 3$ and $3 \times 1 = 3$

$2 \times 3 = 6$ and $3 \times 2 = 6$

$3 \times 3 = 9$ and $3 \times 3 = 9$

$4 \times 3 = 12$ and $3 \times 4 = 12$.

These are both ways of writing the **3 times table**.

6) What numbers are **missing**?

a) ___ × 3 = 9   b) ___ × 3 = 12   c) 3 × ___ = 21

d) ___ × 3 = 15   e) 3 × ___ = 24   f) ___ × 3 = 30

g) 3 × ___ = 27   h) ___ × 3 = 18   i) 0 × 3 = ___

7) A family gets 3 potatoes each for dinner.

There are 5 people in the family.

How many potatoes are used?

8) A table is 2 m long.

3 tables are put in a row to make a big table.

How long is the big table?

★ **9)** Tom, Pip and Dan buy 7 ride tickets each for a funfair.

How many tickets do they buy?

When you **treble** a number, you **multiply by 3**.

Count the counters.

Treble 4 is 12.

4 × 3 = 12

10) In your jotter, copy and complete:

a) **treble** 2 = 2 × 3 = _____

b) **treble** 5 = 5 × 3 = _____

c) **treble** 6 = _____ × _____ = _____

⭐ d) **treble** _____ = _____ × 3 = 30

⭐ e) **treble** _____ = _____ × _____ = 9

Let's try this!

Work in pairs.

Make arrays with counters to show the **2 times table** and **3 times table**.

Which **multiplications** are in both the **2 times** table and the **3 times** table?

Revisit, review, revise

1) In your jotter, copy and complete:

a) 2 × 3 = _____

b) 2 × 6 = _____

c) 2 × _____ = 8

d) 2 × _____ = 20

e) _____ × 2 = 4

f) _____ × 2 = 10

g) 2 × _____ = 2

h) 2 × _____ = 14

i) 2 × 9 = _____

j) _____ × 2 = 16

k) 2 × _____ = 0

→

2) In your jotter, copy and complete:

a) $10 \times 3 = $ _____

b) $2 \times 3 = $ _____

c) $3 \times $ _____ $= 12$

d) $3 \times $ _____ $= 30$

e) _____ $\times 3 = 9$

f) _____ $\times 3 = 3$

g) $3 \times $ _____ $= 15$

h) $3 \times $ _____ $= 18$

i) $3 \times 8 = $ _____

j) _____ $\times 3 = 27$

k) $3 \times $ _____ $= 0$

3) In your jotter, write a **multiplication** that is the same as:

a) double 7

b) double 9

c) treble 2

d) treble 5.

4) In your jotter, work out these **doubles** and **trebles**:

a) double 2

b) double 8

c) treble 3

d) treble 10

9 Measurement
Estimating and measuring

Lengths to the nearest centimetre

> 💡 I will learn to estimate and measure lengths to the nearest centimetre.

The line PQ below has a length that is nearer to 4 cm than to 5 cm.

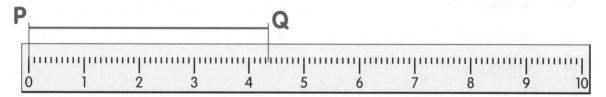

Line PQ is approximately 4 cm, or PQ is 4 cm (to the nearest cm).

Exercise 1

1) Estimate (guess) the length of each line below.
   Give your answers to the nearest centimetre.

   a)

   b)

   c)

   d)

   e)

   f)

2) Use your ruler to measure each of the lines in question 1. Give your answers to the nearest centimetre.

3) **a)** Write the lines in question 1 in order of length, longest first.

b) Find the difference between the shortest and the longest lines.

4) Estimate (guess) the length of each picture below. Give your answers to the nearest centimetre.

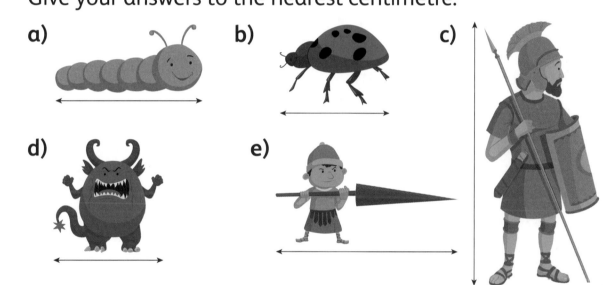

a)

b)

c)

d)

e)

5) Use your ruler to measure each picture in question 4. Give your answers to the nearest centimetre.

⭐ **6)** **a)** Measure each side of this shape to the nearest centimetre.

b) Calculate the difference between the longest and the shortest sides.

Metres, centimetres and millimetres

 I will learn to convert between metres, centimetres and millimetres.

1 metre is equal to **100 centimetres**.

We write **1 m** = **100 cm**.

2 metres is equal to 2 times 100 = 200 cm.

We write **2 m** = **200 cm**.

1 centimetre is equal to **10 millimetres**.

We write **1 cm** = **10 mm**.

Here is a 7 cm line.

7 cm = 70 mm.

Exercise 2

 1) How many **centimetres** (cm) are in each length?

   a) 1 m   **b)** 5 m   **c)** 8 m

   d) 3 m   **e)** 9 m   **f)** 10 m

 2) How many **millimetres** (mm) are in each length?

   a) 1 cm   **b)** 7 cm   **c)** 8 cm

   d) 10 cm   **e)** 9 cm   **f)** 4 cm

⭐ **3)** How many **metres** (m) is:

  a) 400 cm **b)** 700 cm **c)** 800 cm

  d) 1000 cm **e)** 200 cm **f)** 100 cm?

⭐ **4)** How many **centimetres** (cm) is:

  a) 20 mm **b)** 100 mm **c)** 80 mm?

This door is 1 m and 98 cm tall.

We can also say the door is 198 cm tall.

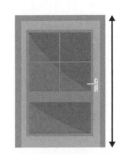

1 m 98 cm = 198 cm

21 cm 5 mm = 215 mm

This book is 9 cm and 5 mm wide.
We can also say the book is 95 mm wide.

5) In your jotter, write the lengths of these lines in mm.

  a)
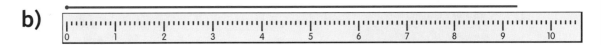

  b)
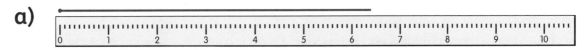

  c)

  d)

⭐ **6)** In your jotter, write the heights of these objects in cm.

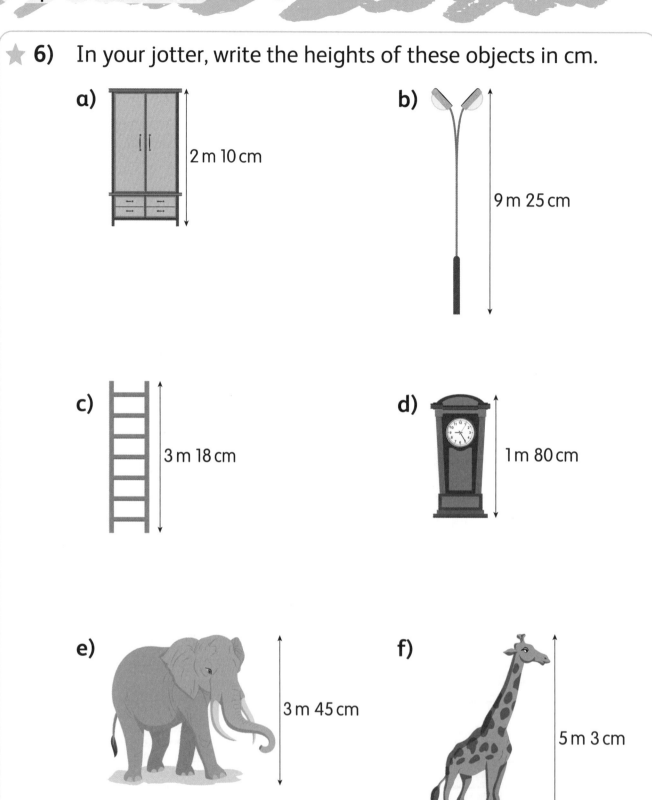

a) 2 m 10 cm

b) 9 m 25 cm

c) 3 m 18 cm

d) 1 m 80 cm

e) 3 m 45 cm

f) 5 m 3 cm

Choosing units of length

 I will learn to choose an appropriate unit of measurement.

Centimetres (cm)

A ruler is used to measure centimetres or millimetres.

Centimetres are used to measure short lengths, such as:

- the length of your finger
- the width of this page.

1 cm = 10 mm

Millimetres are used to measure very short lengths.

- The width of the wire of a paperclip is about 1 mm.
- The width of your fingernail is about 6 mm.

Metres (m)

A metre stick, a tape or a trundle wheel is used to measure metres.

Metres are used to measure longer lengths, such as:

- the length of the classroom
- the height of a building.

1 metre = 100 cm

Kilometres (km)

An odometer on a car is used to measure kilometres.

Kilometres are used to measure much longer distances, such as:

- the length of a river

- the distance from Ayr to Glasgow.

1 km = 1000 m

Exercise 3

1) In your jotter, write which unit of measurement (**mm, cm, m** or **km**) you would use to measure:

a) the width of this page

b) the length of Scotland

c) the width of the school corridor

d) the width of a pencil

e) the length of a pencil

f) the thickness of a ruler

g) the distance from London to Paris

h) the height of a flagpole?

2) What device would you use to measure something in:

    a) millimetres (mm)

    b) centimetres (cm)

    c) metres (m)

    d) kilometres (km)?

3) In your jotter, write things that you might measure in:

    a) millimetres (mm)

    b) centimetres (cm)

    c) metres (m)

    d) kilometres (km).

★ 4) Jack measures the distance from his house to the school using a ruler.
His sister Ann measures the same distance using a trundle wheel.

    a) Which tool is better?   **b)** Why?

Let's try this!

In your jotter, write a list of five things you can measure in the classroom.

What will you use to measure each one?

With a partner, measure your five things.

Area

💡 **I will learn to find the area of a shape by counting square centimetres.**

The area of a shape is the amount of space the shape covers.

A square measuring 1 cm by 1 cm has an area of 1 square centimetre.

This is written as 1 cm².

This shape is made up of 4 squares that measure 1 cm by 1 cm.

It has an area of 4 cm².

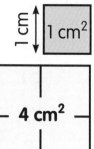

Exercise 4

1) In your jotter, write the area (_____ cm²) of this shape:

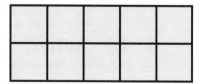

2) In your jotter, write the area (_____ cm²) of these shapes:

a)

b)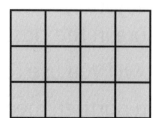

c)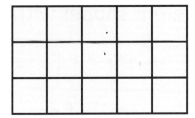

d)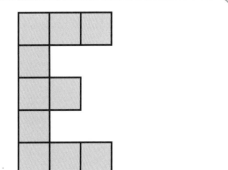

e)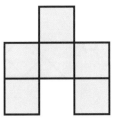

f)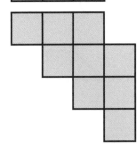

g)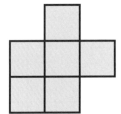

h)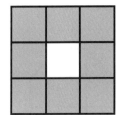

i)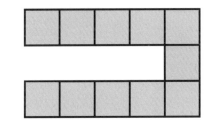

3) On squared paper, draw a rectangle with area 20 cm².

4) In your jotter, write the area of each shape (____ cm²).

a) 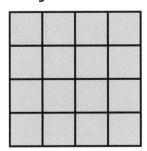   b)

c) What do you notice?

⭐ **5)** On squared paper, draw 5 different shapes with an area of 6 cm².

⭐ **6)** In your jotter, write the area (___ cm²) of each shape.

 $= \dfrac{1}{2}$ cm²

a)

b)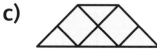

c)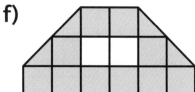

d)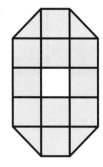

e)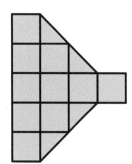

f)

g)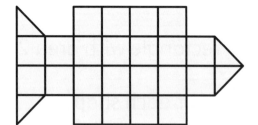

h)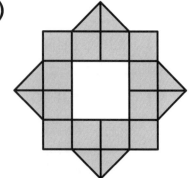

Estimating area

💡 **I will learn to estimate the area of a shape by counting square centimetres.**

It is possible to estimate the area of a shape which does not have straight sides.

Example

To find the area of this blue shape:

- Begin by counting all the whole squares.

- Add on any bits that cover more than half a square.

- Ignore any bits that cover less than half a square.

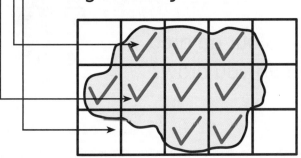

An estimate for the area of this shape is **9 cm²**.

Exercise 5

 1) Estimate the area (_____ cm²) of this shape:

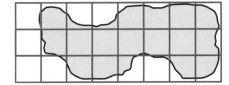

2) Estimate the area of each shape.

a)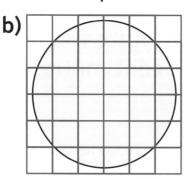

b)

c)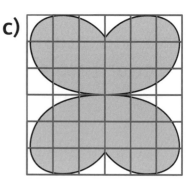

d)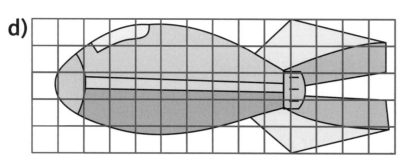

Let's try this!

How many different ways can you colour an area of 24 cm²? What different shapes can you draw? Colour on squared paper.

Revisit, review, revise

1) **a)** Estimate the length of each line:

A _____

B _____

C _____

D _____

b) Measure the length of each line. In your jotter, write each length to the nearest centimetre.

c) In your jotter, write the lengths of the lines in order of size – shortest first.

2) In your jotter, write the unit (**mm**, **cm**, **m** or **km**) you would use to measure:

a) the width of the playground

b) the width of your pencil

c) your height

d) the distance from London to Glasgow.

3) What would you use to measure the lengths in question 2?

(Choose from **ruler**, **metre stick**, **trundle wheel** or **odometer**.)

4) In your jotter, copy and complete:

a) 2 m = ___ cm

b) 4 cm = ___ mm

c) 1 m 17 cm = ___ cm

d) 3 cm 4 mm = ___ mm

e) 3 m 8 cm = ___ cm

5) In your jotter, draw:

a) a rectangle with length 8 cm and width 3 cm

b) a square with side length 7 cm

c) a rectangle with length 5 cm and width $2\frac{1}{2}$ cm.

6) Which has a smaller area:

    a) this page or a stamp

    b) the floor or the window?

7) In your jotter, write the area of each shape in cm².

    a) 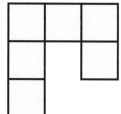   **b)** 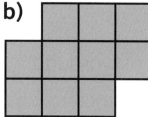   **c)**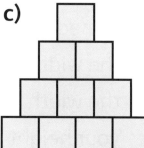

    d) 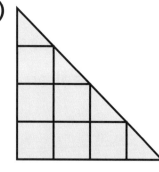   **e)** 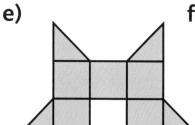   **f)**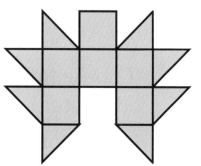

8) Estimate the area of each shape below.

    a) 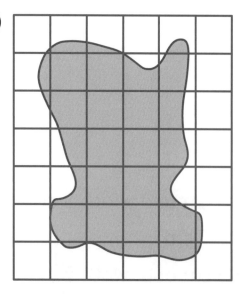   **b)**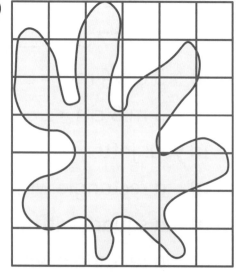

10 Whole numbers 6
Dividing by 2 and 3 and other calculations

Dividing by 2

 I will learn to divide by 2.

Dividing by 2 is the same as **sharing equally between 2**.

Freddie has 2 sweets.

He **shares them equally** with Joe.

Freddie and Joe each get 1 sweet.

We say 2 ÷ 2 = 1.   2 **divided by 2** is 1.

Asha has 4 sweets.

She **shares them equally** with Ruby.

Asha and Ruby each get 2 sweets.

We say 4 ÷ 2 = 2.   4 **divided by** 2 is 2.

Tommy has 6 sweets.

He **shares them equally** with Maisie.

Tommy and Maisie each get 3 sweets.

We say 6 ÷ 2 = 3.   6 **divided by** 2 is 3.

Exercise 1

You will need cubes or counters.

Work with a friend.

1) **a)** Count 8 cubes or counters.

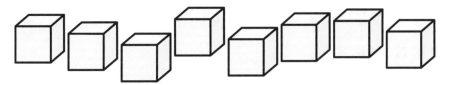

b) **Share them equally** between you and your friend.

How many do you each get?

c) In your jotter, copy and complete:

8 divided by 2 = ____ or 8 ÷ 2 = ____

2) **a)** Count 10 cubes or counters.

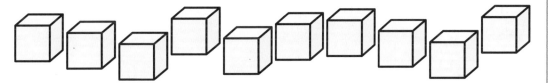

b) **Share them equally** between you and your friend.

How many do you each get?

c) In your jotter, copy and complete:

10 divided by 2 = ____ or 10 ÷ 2 = ____

3) **a)** Count 14 cubes or counters.

Share them equally between you and your friend.

b) In your jotter, copy and complete:

14 divided by 2 = ____ or 14 ÷ 2 = ____

4) In your jotter, copy and complete.
Use cubes or counters to help you.

a) $12 \div 2 =$ _____   b) $16 \div 2 =$ _____

Dividing by 2 is the same as **equal grouping** into 2s.

Freddie has 8 sweets.

He groups them in bags of 2 sweets.

Each time he puts 2 sweets in a bag, there are 2 fewer sweets.

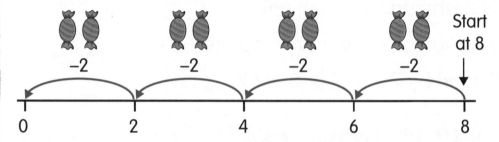

2 into 8 goes 4 times. $8 \div 2 = 4$

---

5) Count back in 2s on these number lines to work out:

a) $10 \div 2 =$ _____

0 1 2 3 4 5 6 7 8 9 10

b) $18 \div 2 =$ _____

0 1 2 3 4 5 6 7 8 9 10 11 12 13 14 15 16 17 18

---

6) In your jotter, copy and complete.
Use cubes, counters or number lines to help you.

a) $2 \div 2$ b) $4 \div 2$ c) $12 \div 2$

Multiplication and division are related.

2 lots of 4 counters = 8 counters.   2 × 4 = 8

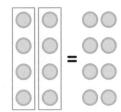

8 counters **shared equally** between 2 = 4.   8 ÷ 2 = 4

4 lots of 2 counters = 8 counters.   4 × 2 = 8

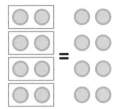

8 counters make 4 **equal groups** of 2.   8 ÷ 2 = 4

2 into 8 goes 4 times.

---

7)   In your jotter, copy and complete the multiplications and the division for each array.

a)   2 × ____ = 6

____ × 2 = 6

6 ÷ 2 = ____

b)   2 × ____ = 10

____ × 2 = 10

10 ÷ 2 = ____

8) In your jotter, copy and complete these multiplications and divisions.
    You may use counters to help you.

    a) $2 \times \underline{\quad} = 20$

    $\underline{\quad} \times 2 = 20$

    $20 \div 2 = \underline{\quad}$

    b) $2 \times \underline{\quad} = 12$

    $\underline{\quad} \times 2 = 12$

    $12 \div 2 = \underline{\quad}$

    c) $2 \times \underline{\quad} = 16$

    $\underline{\quad} \times 2 = 16$

    $16 \div 2 = \underline{\quad}$

    d) $2 \times \underline{\quad} = 14$

    $\underline{\quad} \times 2 = 14$

    $14 \div 2 = \underline{\quad}$

★ 9) In your jotter, write **two multiplications and a division** with an answer of:

    a) 6

    b) 18.

Exercise 2

1) In your jotter, copy and complete:

    a) $8 \div 2$

    b) $6 \div 2$

    c) $10 \div 2$

    d) $14 \div 2$

    e) $18 \div 2$

    f) $16 \div 2$

    g) $20 \div 2$

    h) $2 \div 2$

    i) $0 \div 2$

2) What numbers are **missing**?

    a) $\underline{\quad} \div 2 = 3$

    b) $\underline{\quad} \div 2 = 6$

    c) $\underline{\quad} \div 2 = 9$

    d) $\underline{\quad} \div 2 = 8$

    e) $\underline{\quad} \div 2 = 7$

    f) $\underline{\quad} \div 2 = 10$

3) 8 lollies are **shared equally** between 2 children.

How many lollies does each child get?

4) 12 carrots are **shared equally** between 2 rabbits.

How many carrots does each rabbit get?

5) Sally makes cupcakes.
She has 14 raspberries.
She puts 2 raspberries on each cupcake.

How many cupcakes does she make?

6) Six muffins are **shared equally** between
Toki and Hannah.

How many muffins do they each get?

7) 20 one pence coins are put in piles of 2.

How many piles are there?

Methods of dividing

 I will learn to divide by 2 using a standard method.

This is how you have been writing divisions:

There is another way to write divisions.

⌐ can be used for writing divisions.

$6 \div 2 = 3$
$8 \div 2 = 4$
$4 \div 2 = 2$

6 divided by 2 is 3
6 shared equally
  by 2 is 3
2 into 6 goes 3 times

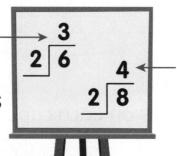

8 divided by 2 is 4
8 shared equally
  by 2 is 4
2 into 8 goes 4 times

Dividing into larger numbers has more steps.

$86 \div 2$ is written:

2 goes into
8 (tens)
4 times

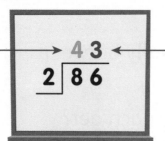

2 goes into
6 (ones)
3 times

$86 \div 2 = 43$

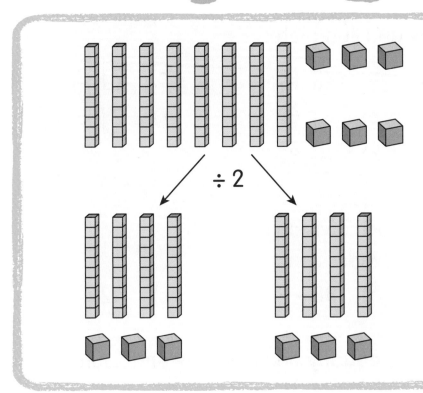

Example

Work out $48 \div 2$:

2 goes into ⟶ **2 4** ⟵ **2** goes into
4 (tens)   **2** | **4 8**   **8** (ones) $48 \div 2 = 24$
2 times   **4** times

Exercise 3

1) What are the **missing** numbers?
   Use cubes or counters to help you.

   a) 2 goes into   **b)** 2 goes into
        2 (tens) ____ times.   8 (ones) ____ times.

              c) ☐☐
                2 | 2 8

2) In your jotter, copy and complete:

a) 2 | 26   b) 2 | 46   c) 2 | 22

d) 2 | 44   e) 2 | 62   f) 2 | 68

g) 2 | 84   h) 2 | 82

3) In your jotter, write each division using $\overline{|}$ then work out the answer.

a) $64 \div 2$ b) $88 \div 2$ c) $42 \div 2$

4) 46 pencils are **shared equally** between Lilly and Laura.

How many pencils does each get?

5) 80 gold coins are **shared equally** between 2 pirates.

How many gold coins does each get?

6) There are 40 children on the sports field. Divide them into teams of 2.

How many teams?

⭐ **7)** Mary sends her mum 48 texts over Saturday and Sunday. She sends the **same** number of texts each day.

How many texts does Mary send on Sunday?

Dividing by 3

 I will learn to divide by 3.

Dividing by 3 is the same as **sharing equally** between 3.

Flo has 3 apples.

She **shares them equally** with Jen and Anna.

Flo, Jen and Anna each get 1 apple.

We say 3 ÷ 3 = 1.   3 **divided by** 3 is 1.

Ben has 6 buns.

He **shares them equally** with Kayo and Isaac.

Ben, Kayo and Isaac each get 2 buns.

We say 6 ÷ 3 = 2.   6 **divided by** 3 is 2.

Connor has 9 biscuits.

He **shares them equally** with Summer and Lina.

Connor, Summer and Lina each get 3 biscuits.

We say 9 ÷ 3 = 3.   9 **divided by** 3 is 3.

Exercise 4

You will need counters or cubes.

You may work in groups of three or on your own.

1) **a)** Count 12 counters or cubes.

b) **Share them equally** among you and 2 friends.

How many do you each get?

c) In your jotter, copy and complete:
12 divided by 3 = _____ or 12 ÷ 3 = _____

2) a) Count 15 counters or cubes.

   b) **Share them equally** among you and 2 friends.

   How many do you each get?

   c) In your jotter, copy and complete:

   15 divided by 3 = ____ or 15 ÷ 3 = ____

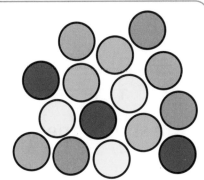

3) a) Count 30 counters or cubes.
   Share them equally among you and 2 friends.

   b) In your jotter, copy and complete:

   30 divided by 3 = ____ or 30 ÷ 3 = ____

4) **Share equally** 24 counters between 3 people.
   In your jotter, copy and complete:
   24 ÷ 3 = ____

Dividing by 3 is the same as **equal grouping** into 3s.

Bob has 12 pears.

He groups them in boxes of 3 pears.

Each time he puts 3 pears in a box, there are 3 fewer pears.

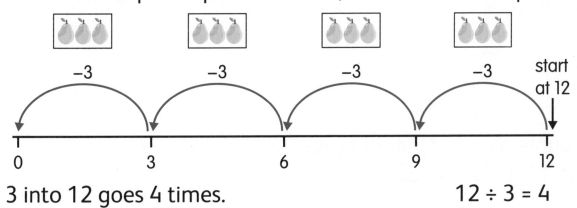

3 into 12 goes 4 times.   12 ÷ 3 = 4

5) Count back in 3s on these number lines to work out:

    a) 15 ÷ 3 = ____

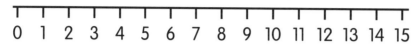

0 1 2 3 4 5 6 7 8 9 10 11 12 13 14 15

    b) 21 ÷ 3 = ____

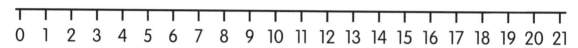

0 1 2 3 4 5 6 7 8 9 10 11 12 13 14 15 16 17 18 19 20 21

6) In your jotter, copy and complete.
Use counters, cubes or number lines to help you.

    a) 9 ÷ 3   **b)** 6 ÷ 3   **c)** 18 ÷ 3

Multiplication and division are related.

3 lots of 4 counters = 12 counters. $3 × 4 = 12$

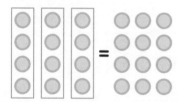

12 counters **shared equally** among 3 = 4. $12 ÷ 3 = 4$

4 lots of 3 counters = 12 counters. $4 × 3 = 12$

12 counters make 4 **equal groups** of 3. $12 ÷ 4 = 3$
3 into 12 goes 4 times.

7) In your jotter, copy and complete the multiplications and the division for each array.

a)   3 × ___ = 6
     ___ × 3 = 6
     6 ÷ 3 = ___

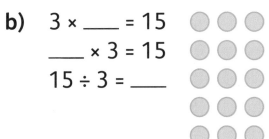

b)   3 × ___ = 15
     ___ × 3 = 15
     15 ÷ 3 = ___

8) In your jotter, copy and complete these.
You may use counters to help you.

a)   3 × ___ = 30
     ___ × 3 = 30
     30 ÷ 3 = ___

b)   3 × ___ = 18
     ___ × 3 = 18
     18 ÷ 3 = ___

c)   3 × ___ = 24
     ___ × 3 = 24
     24 ÷ 3 = ___

d)   3 × ___ = 9
     ___ × 3 = 9
     9 ÷ 3 = ___

9) In your jotter, write **two multiplications and a division** with an answer of:

a)   21   b)   27.

Exercise 5

1) In your jotter, copy and complete:

   a) $3 \div 3$ 　　b) $15 \div 3$ 　　c) $12 \div 3$

   d) $24 \div 3$ 　　e) $21 \div 3$ 　　f) $27 \div 3$

   g) $30 \div 3$ 　　h) $0 \div 3$ 　　i) $18 \div 3$

2) What numbers are **missing**?

   a) ___ $\div 3 = 7$ 　　b) ___ $\div 3 = 3$

   c) ___ $\div 3 = 6$ 　　d) ___ $\div 3 = 5$

   e) ___ $\div 3 = 8$ 　　f) ___ $\div 3 = 10$

3) 6 plants are **shared equally** among 3 gardeners.

   How many plants does each gardener get?

4) Dad gives 27 chips to his 3 children.
   Each child gets the same number of chips.

   How many chips does each child get?

5) I put 3 oranges on each plate.
   I have 12 oranges.

   How many plates do I have?

⭐ 6) A menu gives a choice of 30 items of food.
   There are equal numbers of starters, mains and desserts.

   How many of each are there?

Dividing into larger numbers has more steps.

96 ÷ 3 is written:

3 goes into
9 (tens)
3 times

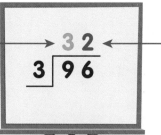

3 goes into
6 (ones)
2 times

96 ÷ 3 = 32

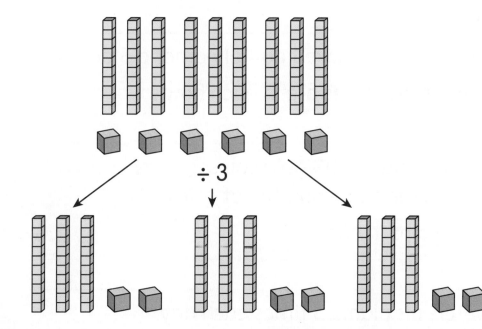

÷ 3

---

7) In your jotter, copy and complete:

a) 3 ⟌ 39

b) 3 ⟌ 66

c) 3 ⟌ 30

d) 3 ⟌ 69

8) In your jotter, write each division using ⌐ then work out the answer.

a) $36 \div 3$

b) $90 \div 3$

c) $96 \div 3$

d) $63 \div 3$

e) $60 \div 3$

f) $33 \div 3$

9) 36 strawberries are **shared equally** among 3 bowls.

How many strawberries are in each bowl?

10) 3 pots have a total of 99 gold coins.
Each pot has the same number of coins.

How many coins in each pot?

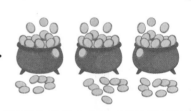

⭐ **11)** Brian waters his plants 3 times each day.
He has watered the plants 96 times in total.

How many days has he been watering his plants?

Missing numbers

 I will learn to solve simple equations involving multiplying and dividing.

Example 1

▱ × 3 = 18

What number times
3 equals 18?

6 × 3 = 18

Answer: ▱ = 6

Example 2

◯ ÷ 2 = 7

What number divided by
2 equals 7?

14 ÷ 2 = 7

Answer: ◯ = 14

Exercise 6

 1) What number does ⬓ stand for in each?

a) ⬓ × 2 = 8

b) ⬓ × 3 = 15

c) ⬓ × 2 = 20

d) ⬓ × 3 = 30

e) ⬓ × 3 = 27

f) ⬓ × 2 = 16

2) What number does ⬭ stand for?

a) ⬭ ÷ 3 = 6

b) ⬭ ÷ 2 = 9

c) ⬭ ÷ 3 = 10

d) ⬭ ÷ 2 = 7

e) ⬭ ÷ 3 = 4

f) ⬭ ÷ 2 = 14

3) What number does ◯ stand for?

a) 3 × ◯ = 6

b) 2 × ◯ = 20

c) 16 ÷ ◯ = 8

d) 3 ÷ ◯ = 1

e) ◯ × 2 = 6

f) ◯ × 3 = 21

g) ◯ ÷ 3 = 5

h) ◯ ÷ 2 = 10

Which sign is missing?

 I will learn to use +, −, × or ÷ to make an equation correct.

Each circle ◯ is covering one of these signs: +, −, × or ÷.

| 6 ◯ 3 = 9 | 8 ◯ 7 = 1 | 5 ◯ 3 = 15 | 8 ◯ 2 = 4 |
| ◯ is + | ◯ is − | ◯ is × | ◯ is ÷ |

Exercise 7

1) What sign does ⬤ stand for?
For each one, try + or −.

    a) 5 ⬤ 5 = 10   **b)** 8 ⬤ 8 = 0

2) What sign does ⬤ stand for?
For each one, try × or ÷.

    a) 2 ⬤ 7 = 14   **b)** 6 ⬤ 3 = 2

3) What sign does ⬤ stand for?
Choose from +, −, × or ÷.

    a) 5 ⬤ 3 = 15   **b)** 9 ⬤ 2 = 7

    c) 16 ⬤ 2 = 8   **d)** 9 ⬤ 7 = 16

    e) 10 ⬤ 2 = 20   **f)** 15 ⬤ 5 = 20

4) In your jotter, copy with +, −, × or ÷ to make each equation true.

    a) 3 ___ 3 = 0   **b)** 7 ___ 3 = 10

    c) 3 ___ 4 = 12   **d)** 21 ___ 3 = 7

    e) 50 ___ 5 = 45   **f)** 7 ___ 8 = 15

    g) 3 ___ 3 = 1   **h)** 3 ___ 3 = 6

    i) 3 ___ 3 = 9   **j)** 18 ___ 3 = 6

5) Roshan is working on this equation, but the second number is covered with ink.

$$4 \rule{1.5em}{0.4pt} \text{✹} = 20$$

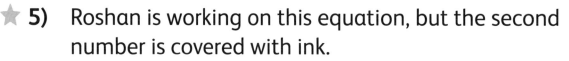

a) In your jotter, copy with a sign and a number to make the equation true.

b) Roshan says there is more than one possible equation. Find another way to make the equation true.

Let's try this!

You will need a pencil and an eraser.

In your jotter, write four equations with an answer of 5.

Each one must have a different sign: +, −, × or ÷.

Now rub out the sign.

For example, Tim's first two equations are:

$4 \rule{1.5em}{0.4pt} 1 = 5$

$5 \rule{1.5em}{0.4pt} 1 = 5$

Swap your four equations with a friend.

Solve each other's equations.

Repeat for the answers 8, 9 and 10.

Revisit, review, revise

1) In your jotter, work out:

    a) $10 \div 2$ **b)** $21 \div 3$ **c)** $18 \div 2$

    d) $18 \div 3$ **e)** $27 \div 3$ **f)** $14 \div 2$

2) In your jotter, work out:

    a) $2\overline{)46}$ **b)** $3\overline{)66}$

3) In your jotter, write each division using $\overline{)}$ then work out the answer.

    a) $64 \div 2$ **b)** $66 \div 2$

    c) $93 \div 3$ **d)** $96 \div 3$

4) 20 sweets are **shared equally** between 2 children.

    How many sweets does each child get?

5) 15 £1 coins are **shared equally** among 3 people.

    How much money does each person get?

6) Ahmed pays 84p for 2 pieces of fruit.
    Each piece of fruit costs the same.

    How much does each piece of fruit cost?

7) 63 raspberries are **shared equally** among 3 people.

    How many raspberries does each person get?

8) What number does $\square$ stand for?

    a) $2 \times \square = 6$ **b)** $3 \times \square = 12$

    c) $\square \times 2 = 20$ **d)** $\square \times 3 = 27$

e) $\boxed{} \div 2 = 6$

f) $27 \div 3 = \boxed{}$

g) $\boxed{} \div 3 = 10$

h) $\boxed{} \div 2 = 0$

9) In your jotter, copy with +, −, × or ÷ to make each equation true.

a) $8 \underline{\hphantom{xx}} 8 = 0$

b) $5 \underline{\hphantom{xx}} 3 = 8$

c) $2 \underline{\hphantom{xx}} 5 = 10$

d) $15 \underline{\hphantom{xx}} 3 = 5$

e) $17 \underline{\hphantom{xx}} 8 = 25$

f) $19 \underline{\hphantom{xx}} 11 = 8$

g) $7 \underline{\hphantom{xx}} 2 = 14$

h) $8 \underline{\hphantom{xx}} 3 = 24$

i) $3 \underline{\hphantom{xx}} 3 = 1$

j) $50 \underline{\hphantom{xx}} 20 = 30$

k) $77 \underline{\hphantom{xx}} 23 = 100$

l) $90 \underline{\hphantom{xx}} 3 = 30$

10) Here are 6 calculations:

8×2 $\qquad\qquad$ $30 \div 3$

$24 - 14$ $\qquad\qquad$ 3×7

$10 + 11$ $\qquad\qquad$ $13 + 3$

Put them into **3 pairs** with an = sign between them.

11 Fractions
Fractions of amounts

Half of an amount

 I will learn how to find halves of amounts.

Finding **half** of an amount is the same as **sharing equally between 2** or **dividing by 2**.

Anna has 12 buttons.

She **shares them equally** with Ben.

They each get **half** the buttons.

Half of 12 is 6.

12 divided by 2 is 6.

$\frac{1}{2}$ of 12 = 6

12 ÷ 2 = 6

Exercise 1

 1) There are 4 pencils in a packet.

How many is **half** of them?

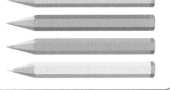

2) There are 20 lollipops in a box.
How many is **half** of them?

3) There are 16 golf balls in a bucket.
How many is **half** of them?

4) **a)** What is **half** of 6?   **b)** What is **half** of 8?

     c) What is **half** of 14?

5) What number is **halfway** between:

a) 0 and 10

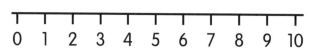

```
0  1  2  3  4  5  6  7  8  9  10
```

b) 0 and 18

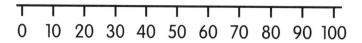

```
0  1  2  3  4  5  6  7  8  9  10  11  12  13  14  15  16  17  18
```

⭐ **c)** 0 and 100?

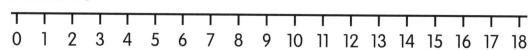

```
0   10  20  30  40  50  60  70  80  90  100
```

⭐ **6)** What number is **halfway** between:

a) 20 and 30

b) 100 and 200

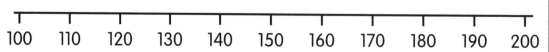

```
100   110   120   130   140   150   160   170   180   190   200
```

c) 200 and 260?

Quarter of an amount

 I will learn how to find quarters of amounts.

Finding a **quarter** of an amount is the same as **sharing equally between 4** or **dividing by 2** and then **dividing by 2** again.

Anna has 12 buttons. She **shares them equally** with Ben.

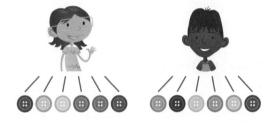

Then Anna and Ben **share equally** with one friend each. They each get a quarter of the buttons.

A **quarter** of 12 is 3.

12 **divided by 2** is 6.

6 **divided by 2** is 3.

$\dfrac{1}{4}$ of 12 = 3

12 ÷ 2 = 6

6 ÷ 2 = 3

Exercise 2

 1) There are 4 pens in a box.

 a) How many is **half** of them?

 b) How many is a **quarter** of them?

2) There are 8 beads on a necklace.
How many is a **quarter** of them?

3) There are 24 chocolates in a box
How many is a **quarter** of them?

⭐ 4) What is a **quarter** of 28?

⭐ 5) What number is a quarter of the way between:

a) 0 and 40

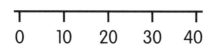

b) 0 and 100?

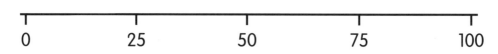

- -

Let's try this!

You will need paper and counters.
Work in pairs.

Imagine:

- the piece of paper is a cake

- the counters are strawberries on the cake.

Put 8 strawberries on the cake.

1) Move the strawberries so that when the cake is cut in **half**, each half has the same number of strawberries on it.

2) Move the strawberries so that when the cake is cut in **quarters**, each quarter has the same number on it.

Repeat with 12 strawberries.

Try with different numbers of strawberries on the cake.

For which numbers is it possible to share the strawberries equally when the cake is cut in:

a) halves   b) quarters?

Revisit, review, revise

1) There are 16 biscuits in a jar.

    a) How many is **half** of them?

    b) How many is a **quarter** of them?

2) There are 24 children in a class.

    a) How many is **half** of them?

    b) How many is a **quarter** of them?

3) a) What number is **halfway** between 0 and 20?

    b) What number is **quarter** of the way between 0 and 20?

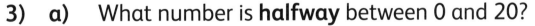
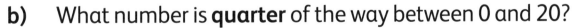
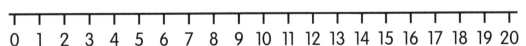

```
0  1  2  3  4  5  6  7  8  9  10  11  12  13  14  15  16  17  18  19  20
```

4) There are 14 t-shirts on a washing line.
Half of them are white. How many are **not** white?

12 Statistics

Graphs, tallies and frequency tables

Reading from a pictograph

💡 I will learn to interpret information from a pictograph.

A **pictograph** is a graph which uses pictures to show information. Each picture stands for a number which can be found in a key.

This pictograph shows the number of Primary 2 children who bring a packed lunch to school.

Key: 🧍 represents 2 children.

Mon	🧍 🧍 🧍 🧍 🧍
Tue	🧍 🧍 🧍 ⸩
Wed	🧍 🧍 ⸩
Thu	🧍 ⸩
Fri	

The pictograph shows 5 figures on Monday.

Each figure represents 2 children.

$5 \times 2 = 10$

10 children brought a packed lunch to school on Monday.

Half a picture stands for $\frac{1}{2}$ the number of children.

$\frac{1}{2}$ of 2 = 1

This half picture represents 1 child.

Exercise 1

 1) This pictograph shows the number of people waiting in a queue for the dodgems at the fun fair.

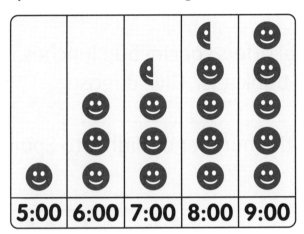

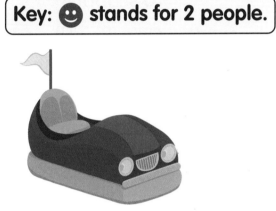

Key: 😊 stands for 2 people.

How many people were in the queue at:

a)   5:00   b)   6:00   c)   7:00

d)   8:00   e)   9:00?

2) This pictograph shows how many children bring a packed lunch to school.

Mon	🧍 🧍 🧍 🧍 🧍
Tue	🧍 🧍 🧍 ⌐
Wed	🧍 🧍 ⌐
Thu	🧍 ⌐
Fri	

Key: 🧍 represents 2 children.

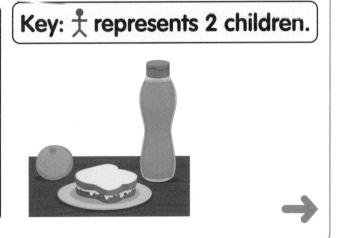

→

a) How many children bring a packed lunch on Tuesday?

b) How many bring a packed lunch on Wednesday?

c) How many children in total bring packed lunches to school?

d) One day, the school offers special 50p lunches. Which day do you think it is? Give a reason.

3) This pictograph shows the number of flights to Spain which an airline makes.

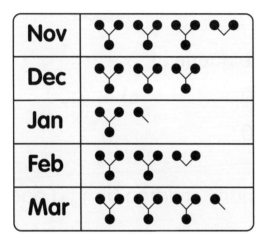

Key: stands for 3 flights.

a) How many flights in December?

b) How many flights in November?

c) Which month has the most flights?

d) Which month has the fewest flights?

e) How many flights in total in February and March?

Drawing pictographs

 I will learn to draw a pictograph.

Example

The table shows the pets the children in a Primary 3 class own.

Cat	Hamster	Dog	Goldfish	Rabbit
6	4	9	3	7

Use the key to draw a **pictograph**.

Key: ☺ **stands for 2 children.**

This means that ◖ stands for 1 child.

Cat	☺ ☺ ☺
Hamster	☺ ☺
Dog	☺ ☺ ☺ ☺ ◖
Goldfish	☺ ◖
Rabbit	☺ ☺ ☺ ◖

Exercise 2

Use the key to draw a pictograph to show the information given in the table.

1) Types of fish for sale.

Cod	Haddock	Plaice	Bass	Sole
5	7	4	2	7

Key: stands for 1 fish.

2) Primary 4's favourite zoo animals.

Key: ☺ **stands for 1 person.**	Lion	Tiger	Giraffe	Seal	Panda
	5	3	4	1	6

3) Number of robins seen in a front garden.

Key: stands for 2 birds.	Mon	Tues	Wed	Thurs	Fri
	4	10	6	8	2

4) Number of people in a queue.

6.30PM	7.00PM	7.30PM	8.00PM	8.30PM
10	12	8	5	3

Key:
🧍 **stands for 2 people.**
♪ = 1 person.

Reading from block graphs and bar graphs

 I will learn to interpret information from a block graph and bar graph.

Block graphs and bar graphs are ways of showing information.

This block graph shows the different vehicles in a car park.

There are 2 bikes in the car park.

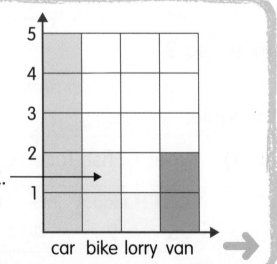

This bar graph shows where people like to go in their free time.

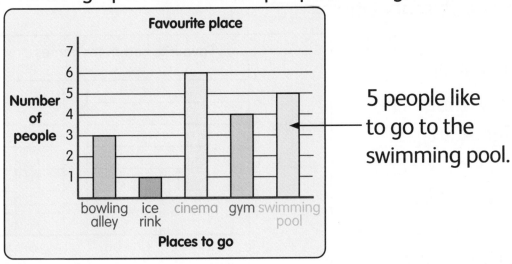

5 people like to go to the swimming pool.

Exercise 3

 1) Look at the **block graph** on page 144 about vehicles in a car park.

    a) How many cars are in the car park?

    b) How many lorries are in the car park?

    c) How many vans are in the car park?

    d) What is the total number of vehicles in the car park?

2) The block graph shows the items sold in a shop one morning.

    a) How many hats were sold?

    b) How many t-shirts were sold?

    c) How many items were sold altogether?

    d) How many **more** t-shirts were sold than socks?

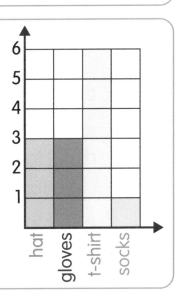

3) The bar graph shows the favourite computer games of the pupils in a Primary class.

a) How many chose Blue Rain?

b) How many chose Red Alert?

c) How many chose Duty Calls?

d) How many chose Cowboy Joe?

e) What is the most popular game?

f) How many pupils were asked altogether?

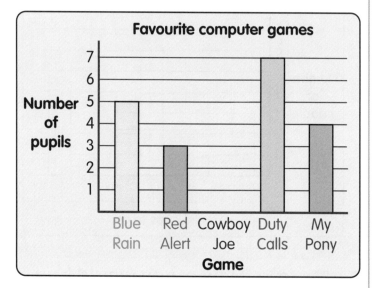

4) The bar graph shows the cakes sold in a bakery.

a) What does the scale on the graph go up in?

b) What is the least popular cake?

c) How many eclairs are sold?

d) How many cupcakes are sold?

e) How many doughnuts are sold?

f) How many more buns are sold than eclairs?

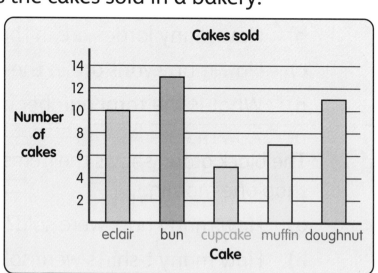

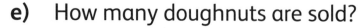

5) People are asked what they would buy if they won the lottery. Their answers are shown in a horizontal bar graph.

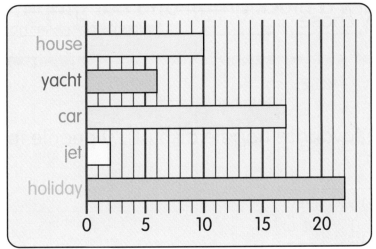

a) How many would buy a house?

b) How many would buy a car?

c) What is the most popular answer? How many voted for that?

d) How many **more** chose yacht than jet?

6) Pupils were asked which musical instrument they like. Here are their answers:

a) How many said harp? (It's not 2!)

b) How many said piano?

c) How many said flute?

d) How many **more** said guitar than violin?

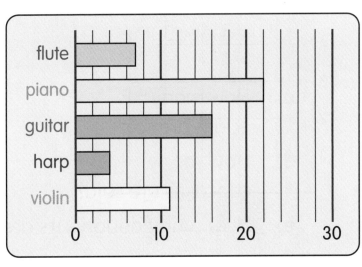

Drawing block graphs and bar graphs

 I will learn to draw a block graph and bar graph.

Example

The table shows the favourite noodle dish of 20 people in a restaurant.

prawn	vegetable	chicken	beef	mixed
5	3	7	1	4

This information can be displayed in a **block graph** or a **bar graph**.

To draw a block graph, colour in the number of squares shown in the table.

Write the types of dish along the bottom.

Write the numbers up the side.

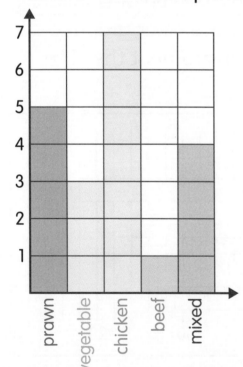

A bar graph needs:

- a title
- a grid of lines
- labels
- bars.

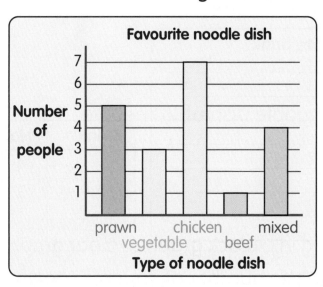

Exercise 4

🚀 **1)** The table shows the hair colour of a group of children. In your jotter, copy and complete the block graph to show the information.

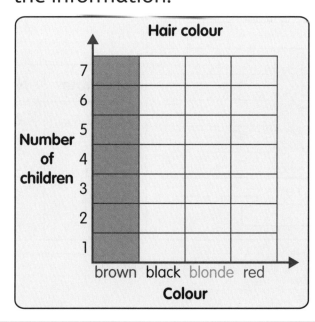

Colour	No.
brown	7
black	2
blonde	4
red	1

2) Twenty children are asked to name their favourite drink.

    a) Draw a block graph to show the information.

    b) Copy and complete the bar graph.

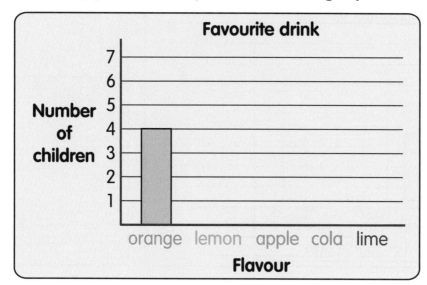

Drink	No.
orange	4
lemon	3
apple	1
cola	7
lime	5

3) Primary 3 pupils were asked to name their favourite colour. In your jotter, copy and complete the bar graph.

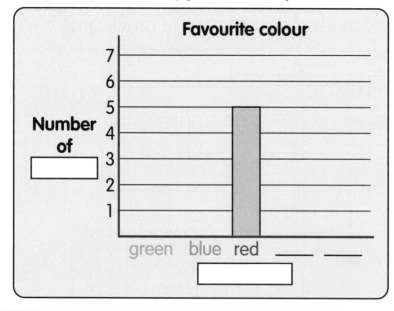

Colour	No.
green	6
blue	7
red	5
pink	2
yellow	4

4) People were asked what creatures they find in their gardens. In your jotter, copy and complete the bar graph.

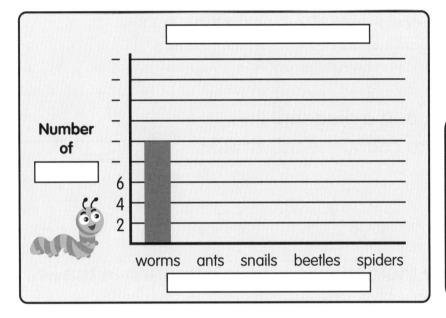

10 said worms

3 said ants

12 said snails

9 said beetles

1 said spiders

Tally marks and frequency tables

💡 **I will learn to use tally marks to count and make frequency tables.**

A **frequency table** makes it easier to understand lists of information.

Example

This list shows the colour of children's eyes.

| blue | brown | green | blue | hazel | blue | green | blue |
| green | brown | brown | hazel | green | grey | blue |

A tally helps to count how many times a value or item occurs.

→

To make a tally, draw a line every time a value or item occurs.

grey 1 = |

hazel 2 = ||

brown 3 = |||

green 4 = ||||

When you get to 5, add a diagonal line.
This makes it easy to count the tally marks.

5 = ⊬⊦⊦

6 = ⊬⊦⊦ |

7 = ⊬⊦⊦ ||

Here is a frequency table showing the information in the list.

Eye colour	Tally	Total
Blue	⊬⊦⊦	5
Brown	|||	3
Green	||||	4
Grey			1
Hazel	||	2

⟵ 5 children have blue eyes.

Exercise 5

1) Look at the frequency table above.

    a) How many children have green eyes?

    b) How many children have brown eyes?

    c) How many children have grey eyes?

    d) What is the most common colour of eyes?

e) How many **more** children have green eyes than hazel eyes?

f) How many **fewer** children have grey eyes than brown eyes?

g) What is the total number of children with blue or brown eyes?

h) How many children were asked **altogether**?

2) People were asked to name any Scottish town starting with an A.

Town	Tally	Total
Ayr	|||	3
Airdrie	⑷		
Annan			

a) In your jotter, copy the table and complete the total column.

b) How many people were asked?

3) People were asked which city they lived nearest to.
In your jotter, copy the table and fill in the gaps.

City	Tally	Total
Dundee	⑷	5
Inverness			
Glasgow		8
Edinburgh		7
Stirling	||	
Perth	||||	

4) Here are the colours of the t-shirts worn on sports day.

| black | orange | red | red | blue | pink | black | black | red | orange |
| red | pink | red | orange | red | pink | red | red | pink | pink |

Colour	Tally	Total
Orange		
Red		
Black		
Blue		
Pink		

a) In your jotter, copy the frequency table and fill in the tally marks and totals.

b) Which colour of t-shirt was worn the most?

c) How many people were asked?

d) How many people did not wear a black t-shirt?

5) What numbers do these tally marks show?

a) |||| |||

b) |||| |||| ||

c) |||| |||| |||| ||||

d) |||| |||| |||| |||| |||| |

e) |||| |||| |||| |||| |||| |||| |||| ||

6) Draw tally marks to show these numbers:

a) 7   b) 10   c) 15

d) 17   e) 20   f) 26

⭐ **7)** Using tally marks, make a frequency table to show the answers to 'what's your favourite flower?'

rose	tulip	pansy	daffodil	pansy
pansy	rose	tulip	pansy	pansy
tulip	pansy	tulip	tulip	daffodil
marigold	tulip	rose	daffodil	marigold
rose	daffodil	tulip	tulip	daffodil
tulip	pansy	tulip	rose	daffodil

Flower	Tally	Total
rose		
pansy		

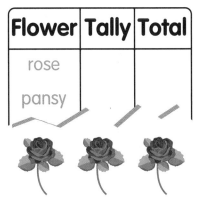

Let's try this!

Work in pairs to carry out a pet survey.

Draw tables to record what pets pupils in your class have.

Pet	Tally	Total
dog		
cat		

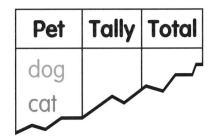

Record your results in a pictograph, block graph and bar graph.

Present your results to the class.

Ask and answer questions about each other's results.

Revisit, review, revise

1) The pictograph shows the number of animals on a small farm.

Key: ◉ stands for 2 farm animals.

Goats	◉ ◉ ◉
Cows	◉ ◉ ◉ ◉
Sheep	◉ ◉ ◉ ◉ ◉
Hens	◉ ◉ ◖
Bulls	◖

a) How many cows?

b) How many hens?

c) How many more sheep than goats?

d) How many bulls?

2) Here is a frequency table showing orders in a restaurant.

a) Copy and complete the frequency table:

Order	Tally	Total
Salad	IIII II	
Soup		4
Omelette		10
Burger	II	

b) Copy and complete the block graph to show the information.

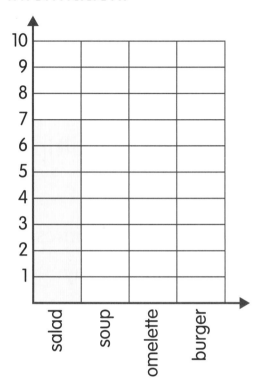

c) In your jotter, copy and complete the bar graph to show the information.

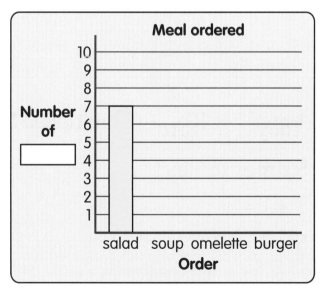

d) How many **more** omelettes than burgers were ordered?

13 Whole numbers 7
Multiplying and dividing by 5 and 10

5 times table

 I will learn the 5 times table and to multiply by 5.

You should know the 2 times table and the 3 times table.

You can learn the **5 times table** in a similar way.

5 lots of 1 counter = 5 counters.

⊡ + ⊡ + ⊡ + ⊡ + ⊡ = ⚪⚪⚪⚪⚪

We say 5 × 1 = 5.　　　5 **times** 1 is 5.　　　5 **multiplied by** 1 is 5.

5 lots of 2 counters = 10 counters.

We say 5 × 2 = 10.　　　5 **times** 2 is 10.　　　5 **multiplied by** 2 is 10.

Here is the start of the **5 times table**:

5 lots of 1 = 5.　　　　5 × 1 = 5

5 lots of 2 = 10.　　　　5 × 2 = 10

Exercise 1

 1) In your jotter, write the title '**5 times table**'.
Now copy and complete:

    a) $5 \times 1 = 5$ **b)** $5 \times 2 =$ ___   **c)** $5 \times 3 =$ ___

    d) $5 \times 4 =$ ___   **e)** $5 \times 5 =$ ___   **f)** $5 \times 6 =$ ___

    g) $5 \times$ ___ $=$ ___   **h)** $5 \times$ ___ $=$ ___   **i)** ___ $\times$ ___ $=$ ___

    j) ___ $\times 10 =$ ___

2) Put tracing paper on top of this grid.

    a) Circle the numbers in the 5 times table.

    b) What do you notice about the last digit of the numbers in the 5 times table?

1	2	3	4	5	6	7	8	9	10
11	12	13	14	15	16	17	18	19	20
21	22	23	24	25	26	27	28	29	30
31	32	33	34	35	36	37	38	39	40
41	42	43	44	45	46	47	48	49	50

3) Ajay is skip counting in 5s on this number line.

He counts 5, 10, 15.

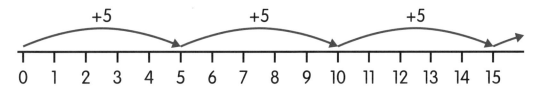

In your jotter, write the next **7 numbers** that he says.

4) Try to learn the **5 times table** by heart.
Say to yourself $5 \times 1 = 5$, $5 \times 2 = 10$, $5 \times 3 = 15$, ...
Now say it aloud to a friend.

1 lot of 5 counters = 5 counters.

We say $1 \times 5 = 5$.
1 **times** 5 is 5.
1 **multiplied by** 5 is 5.

2 lots of 5 counters = 10 counters.

We say $2 \times 5 = 10$.
2 **times** 5 is 10.
2 **multiplied by** 5 is 10.

Notice that:

$1 \times 5 = 5$ and $5 \times 1 = 5$

$2 \times 5 = 10$ and $5 \times 2 = 10$.

These are both ways of writing the **5 times table**.

5) What numbers are **missing**?
   a) $5 \times \underline{\hphantom{xx}} = 25$
   b) $\underline{\hphantom{xx}} \times 5 = 5$
   c) $5 \times \underline{\hphantom{xx}} = 15$
   d) $5 \times \underline{\hphantom{xx}} = 35$
   e) $5 \times \underline{\hphantom{xx}} = 45$
   f) $\underline{\hphantom{xx}} \times 5 = 20$
   g) $\underline{\hphantom{xx}} \times 5 = 30$
   h) $5 \times \underline{\hphantom{xx}} = 40$
   i) $0 \times 5 = \underline{\hphantom{xx}}$

6) There are 5 buns in a pack.
How many buns in 7 packs?

7) Ellie has a pile of 9 five pence coins.
How much money does she have?

Sometimes the answer to a multiplication is called a **product**.
The **product** of 4 and 5 is 20, because 4 × 5 = 20.

 8) What is the **product** of:
a) 2 and 5   b) 5 and 3   c) 2, 3 and 5?

Dividing by 5

 I will learn to divide by 5.

Dividing by 5 is the same as **sharing equally** among 5.
Finley has 5 grapes.

He **shares them equally** with Lola, Pam, Shazia and Alice.

All 5 children each get 1 grape.

We say 5 ÷ 5 = 1.
5 **divided by** 5 is 1.

161

Ruby has 10 grapes.

She **shares them equally** with Mabel, Zak, Ming and Jesse.

All 5 children each get 2 grapes.

We say $10 \div 5 = 2$.

10 **divided by** 5 is 2.

Exercise 2

1) You will need 15 counters and 5 small squares of paper.
Work in pairs.

a) Share the counters equally among the 5 squares
of paper.
How many counters on each square?

b) In your jotter, copy and complete:
15 divided by 5 is ____ or $15 \div 5 =$ ____

2) In your jotter, copy and complete.
Use counters to help you.

a) $1 \times 5 =$ ____ and $5 \div 5 =$ ____

b) $2 \times 5 =$ ____ and $10 \div 5 =$ ____

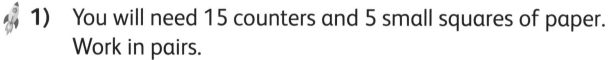

c) 3 × 5 = ___ and 15 ÷ 5 = ___

3) In your jotter, write out your 5 times table.

4) Use your answers to question 3 to copy and complete:

a) 20 ÷ 5 = ___   **b)** 10 ÷ 5 = ___

c) 50 ÷ 5 = ___   **d)** 35 ÷ 5 = ___

e) 25 ÷ 5 = ___   **f)** 40 ÷ 5 = ___

g) 45 ÷ 5 = ___   **h)** 30 ÷ 5 = ___

5) What are the **missing** numbers?

a) ___ ÷ 5 = 3   **b)** ___ ÷ 5 = 7

c) ___ ÷ 5 = 10   **d)** ___ ÷ 5 = 6

e) ___ ÷ 5 = 8   **f)** ___ ÷ 5 = 5

6) In your jotter, work out:

a) How many £5 notes are equal to £45?

b) How many 5p coins can you get for 20p?

c) 40 children are in a sports club.
They are grouped into teams of 5 children.
How many teams do the children make?

d) Buses park in rows of 5.
There are 35 buses.
How many rows in total?

    e)   There are 5 frying pans and 30 eggs.
Eggs are divided equally among the pans.
How many eggs in each pan?

    f)   25 books are divided equally between 5 shelves.
How many books on each shelf?

⭐ **g)**   Ibrahim pays 25p for 5 chews.
Alfred buys 2 of the same chews.
How much does Alfred pay?

10 times table

 I will learn the 10 times table and to multiply by 10.

You should know the 2 times table, the 3 times table and the 5 times table.

You have met the **10 times table** before, when using ten frames:

10 lots of 1 counter = 10 counters.

We say 10 × 1 = 10.   10 **times** 1 is 10.   10 **multiplied by** 1 is 10.

We can also look at the whole ten frame.

1 lot of 10 counters = 10 counters.

We say 1 × 10 = 10.   1 **times** 10 is 10.   1 **multiplied by** 10 is 10.

Exercise 3

 1) In your jotter, write the title '**10 times table**'.
Use the ten frames to help you copy and complete:

a) 10 × 1 = ____

b) 10 × 2 = ____

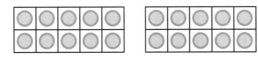

c) 10 × 3 = ____

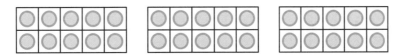

d) 10 × 4 = ____

2) Copy and complete the 10 times table up to 10 × 10:

a) 10 × 5 = ____

b) 10 × 6 = ____

c) 10 × ____ = ____

d) 10 × ____ = ____

e) ____ × ____ = ____

f) ____ × 10 = ____

3) Put tracing paper on top of this grid.

1	2	3	4	5	6	7	8	9	10
11	12	13	14	15	16	17	18	19	20
21	22	23	24	25	26	27	28	29	30
31	32	33	34	35	36	37	38	39	40
41	42	43	44	45	46	47	48	49	50
51	52	53	54	55	56	57	58	59	60
61	62	63	64	65	66	67	68	69	70
71	72	73	74	75	76	77	78	79	80
81	82	83	84	85	86	87	88	89	90
91	92	93	94	95	96	97	98	99	100

a) Circle the numbers in the 10 times table.

b) What do you notice about the last digit of the numbers in the 10 times table?

4) Ethel is skip counting in 10s on this number line.
She counts 10, 20.

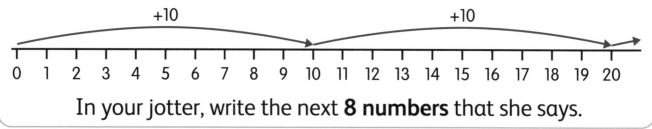

In your jotter, write the next **8 numbers** that she says.

5) Try to learn the **10 times table** by heart.
Say to yourself 10 × 1 = 10, 10 × 2 = 20, 10 × 3 = 30, ...
Now say it aloud to a friend.

6) What numbers are **missing**?

a) 10 × ___ = 50

b) ___ × 10 = 20

c) ___ × 10 = 80

d) 10 × ___ = 30

e) 10 × ___ = 60

f) ___ × 10 = 70

g) 10 × ___ = 90

h) ___ × 10 = 40

i) 10 × ___ = 100

j) 0 × 10 = ___

7) A jar contains 7 ten pence coins.
How much money is in the jar?

8) There are 10 pens in a box.
How many pens in 8 boxes?

9) 10 people each borrow 5 books from the library.
How many books are borrowed?

10) What is the **product** of:

    a)   2 and 10

    b)   10 and 4

    c)   3, 2 and 10

    d)   2, 5 and 10?

Dividing by 10

 I will learn to divide by 10.

Dividing by 10 is the same as **sharing equally** among 10.

10 strawberries are **shared equally** among 10 children.

All 10 children each get 1 strawberry.

We say 10 ÷ 10 = 1.   **10 divided by** 10 is 1.

20 grapes are **shared equally** among 10 children.

All 10 children each get 2 grapes.

We say 20 ÷ 10 = 2.   20 **divided by** 10 is 2.

Exercise 4

1) You will need counters and 10 small squares of paper.
Work in pairs.

a) Count 30 counters.

b) Share the counters equally among the 10 squares
of paper.
How many counters on each square?

c) In your jotter, copy and complete:

30 divided by 10 is ____ or 30 ÷ 10 = ____

2) In your jotter, copy and complete.

Use counters to help you.

a) 1 × 10 = ____ and 10 ÷ 10 = ____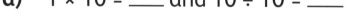

b) 2 × 10 = ____ and 20 ÷ 10 = ____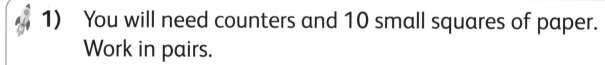

c) 3 × 10 = ____ and 30 ÷ 10 = ____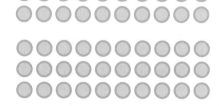

3) In your jotter, write out your 10 times table.

4) Use your answers to question 3 to copy and complete:

a) $50 \div 10 =$ _____

b) $40 \div 10 =$ _____

c) $80 \div 10 =$ _____

d) $70 \div 10 =$ _____

e) $60 \div 10 =$ _____

f) $90 \div 10 =$ _____

5) What are the **missing** numbers?

a) _____ $\div 10 = 3$

b) _____ $\div 10 = 5$

c) _____ $\div 10 = 9$

d) _____ $\div 10 = 8$

e) _____ $\div 10 = 6$

f) _____ $\div 10 = 10$

6) In your jotter, work out:

a) **i)** How many £10 notes do you get for £70?

ii) How many 10p coins do you get for 50 pence?

b) Flowers are grouped in 10s to make bunches.
There are 80 flowers.
How many bunches are there?

c) A pet shop sells dog biscuits for 10p each.
Harry pays 60p for dog biscuits.
How many biscuits does he buy?

d) 30 balloons are shared equally among 10 children.
How many balloons does each child get?

e) Juice is sold in trays of 10 cartons.
A shopper wants 100 cartons.
How many trays must they buy?

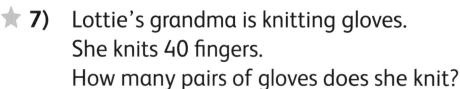

⭐ **7)** Lottie's grandma is knitting gloves.
She knits 40 fingers.
How many pairs of gloves does she knit?

Let's try this!

Work in pairs.

You will need two pieces of paper.

On one piece of paper, write a quiz with 5 multiplication questions and 5 division questions.

On the other piece of paper, write the answers.

Now swap quizzes with another pair. Then do each other's quiz.

Swap back to mark and score.

Revisit, review, revise

1) In your jotter, work out:

a) 8 × 5	**b)** 3 × 5
c) 6 × 5	**d)** 5 × 5
e) 5 × 7	**f)** 5 × 1
g) 5 × 9	**h)** 5 × 0
i) 2 × 5	**j)** 5 × 4
k) 10 × 5	**l)** 5 × 3

→

2) In your jotter, find:

a) 7 × 10   **b)** 9 × 10

c) 10 × 3   **d)** 10 × 6

e) 10 × 2   **f)** 4 × 10

g) 1 × 10   **h)** 10 × 0

i) 8 × 10   **j)** 10 × 10

k) 5 × 10   **l)** 10 × 8

3) What are the **missing** numbers?

a) 5 × ___ = 35   **b)** 5 × ___ = 45

c) 10 × ___ = 40   **d)** 10 × ___ = 90

e) ___ × 5 = 15   **f)** ___ × 5 = 20

g) ___ × 10 = 20   **h)** ___ × 5 = 40

4) In your jotter, copy and complete the multiplications and the division for each array.

a) 5 × ___ = 10
    ___ × 5 = 10
    10 ÷ 5 = ___

b) 10 × ___ = 30
    ___ × 10 = 30
    30 ÷ 10 = ___

5) In your jotter, copy and complete these multiplications and divisions.

You may use counters to help you.

a) $10 \times \_\_\_ = 80$

$\_\_\_ \times 10 = 80$

$80 \div 10 = \_\_\_$

b) $5 \times \_\_\_ = 50$

$\_\_\_ \times 5 = 50$

$50 \div 5 = \_\_\_$

6) A builder is building 10 flats.
Each flat has 7 doors.
How many doors does the builder need?

7) Barry saves £10 each month.
He says: 'Now I have saved £90'.
How many months has Barry been saving?

8) An adult's ticket to a football match costs £10.
A child's ticket costs £5.

a) What is the cost for 5 adults?

b) What is the cost for 4 children?

c) What is the cost for 2 adults and 3 children?

14 Time
Digital clocks, measuring time and the calendar

Digital clocks

💡 **I will learn to tell the time on a digital clock.**

There are 60 minutes in one hour.

'Half past' is 30 minutes after the hour.

'Quarter past' is 15 minutes after the hour.

'Quarter to' is 15 minutes before the next hour.

Digital clocks show the time using numbers only.

hour   separator   minutes

8:00   8 o'clock

2:30   half past 2

6:15   quarter past 6

3:45   quarter to 4

45 minutes **past** 3 is the same as 15 minutes **to** 4.

Exercise 1

 1) In your jotter, write the time shown on each clock.
Use the following words: **half past**, **quarter past**, **quarter to** and **o'clock**.

a)
4:30
(half past ...)

b)
5:15
(quarter past ...)

c)
1:45
(quarter to ...)

d)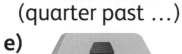
8:30

e)
9:00

f)
6:45

g)
2:30

h)
9:45

i)
11:15

j)
9:30

k)
1:15

l)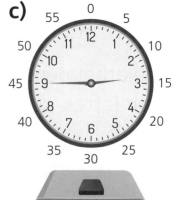
5:45

2) In your jotter, write the digital time for each clock face.

a)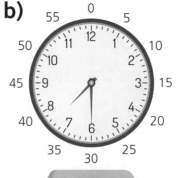
5:....

b)
7:....

c)
....:....

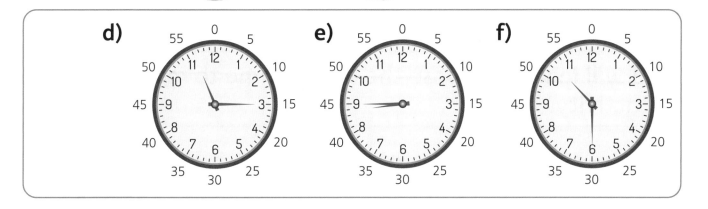

d) **e)** **f)**

3) In your jotter, write the digital time.

  a) half past 9   **b)** quarter past 1

  c) quarter to 9   **d)** quarter past 3

  e) quarter to 12   **f)** half past 2

  g) quarter past 7   **h)** quarter to 11

  i) half past 6   **j)** quarter to six

Let's try this!

Work in pairs.

One person shows the time on a clock face.

Make half past, quarter past, quarter to or o'clock times.

The other person writes the digital time in their jotter.

Take it in turns.

Units of time

> **I will learn to choose the unit of time to measure an activity.**

There are 60 seconds in 1 minute.

There are 60 minutes in 1 hour.

There are 24 hours in 1 day.

Exercise 2

 1) Would you measure in **seconds**, **minutes**, **hours** or **days**?

    a) walking across the classroom

    b) walking all round Scotland

    c) doing a 500-piece jigsaw

    d) writing your name

    e) boiling an egg

    f) watching two films, one after the other

2) List two activities that would take:

    a) seconds   **b)** minutes

    c) hours   **d)** days.

3) Are there **more** or **less** than 24 hours in a weekend?

4) Are there **more** or **less** than 60 minutes in a school day?

5) Does it take **more** or **less** than 60 seconds to take one breath?

6) Match each period of time to the correct activity.

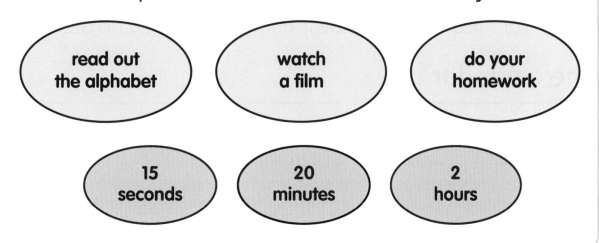

7) Match each period of time to the correct activity.

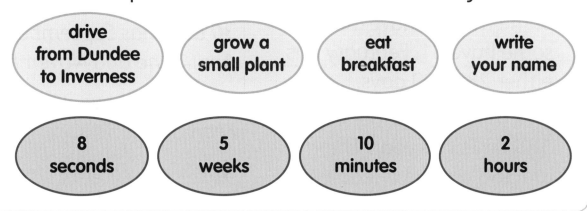

8) Use the internet to find out how long it takes to run:

   a) 100 metres   **b)** 1 mile   **c)** a marathon.

9) Make a poster about one of these topics, or choose your own topic.

Use the internet to find the information.

a) how long it takes to fly to different cities in Europe

b) how many hours of practice different sportspeople do each week

c) how old the oldest animals in the world are

The calendar

 I will learn to use a calendar to answer questions.

There are **7 days** in a **week**.

There are **12 months** in a **year**.

Some months have 30 days, some have 31. February has either 28 or 29 days.

This means there are usually **365 days** in a **year**. There are **366 days** in a **leap year**!

There are approximately **52 weeks** in a **year**.

You can use a poem to remember how many days there are in each month.

30 days has September, April, June and November.

All the rest have 31, except February,

which has 28 days clear, and 29 days each leap year.

Remember, remember

There are 12 months in a year. The order of the months is:

1) January   **2)** February   **3)** March   **4)** April

5) May   **6)** June   **7)** July   **8)** August

9) September   **10)** October   **11)** November   **12)** December

---

The date 3 January 2023 can be written using 6 digits.

3 January 2023 = 03.01.23 or
03/01/23

day   month   year

We write January as 1 because it is the 1st month in the year.

2023		JANUARY			2023	
SUN	MON	TUE	WED	THU	FRI	SAT
1	2	③	4	5	6	7
8	9	10	11	12	13	14
15	16	17	18	19	20	21
22	23	24	25	26	27	28
29	30	31				

Exercise 3

1) In your jotter, write each date using 6 digits.

    a) 23 February 2023   **b)** 19 April 2024

    c) 22 July 2019   **d)** 18 August 2022

    e) 7 June 2014   **f)** 3 March 2030

    g) 10 December 2016   **h)** 4 February 2021

2) In your jotter, write each date in words.

    a) 14/01/24   **b)** 01/03/25   **c)** 11/11/16

    d) 23/04/11   **e)** 12.12.17   **f)** 07.08.20

    g) 09.03.25   **h)** 30.06.21   **i)** 28/02/15

3) Jim writes the date 30.02.23.
What is wrong with this date?

4) Look at the month of April 2025.

    a) How many Wednesdays are in April 2025?

    b) Ellie's birthday is on 03/04/25.
What **day** is her birthday on?

2025		APRIL			2025	
SUN	MON	TUE	WED	THU	FRI	SAT
		1	2	3	4	5
6	7	8	9	10	11	12
13	14	15	16	17	18	19
20	21	22	23	24	25	26
27	28	29	30			

    c) Ayla's birthday is exactly 3 weeks after Ellie's.
What **date** is Ayla's birthday?

    d) Ronnie's birthday is 3 days **after** Ellie's.
What **day** is Ronnie's birthday on?

★ **5)** Look at the two calendar months shown.

    a) How many Saturdays in total in May and June 2026?

    b) My birthday is on 14 May. My sister's birthday is exactly 3 weeks later. What **date** is my sister's birthday?

2026		MAY			2026	
SUN	MON	TUE	WED	THU	FRI	SAT
					1	2
3	4	5	6	7	8	9
10	11	12	13	14	15	16
17	18	19	20	21	22	23
24	25	26	27	28	29	30
31						

c) What is the **day** and **date**:

2026		JUNE			2026	
SUN	MON	TUE	WED	THU	FRI	SAT
	1	2	3	4	5	6
7	8	9	10	11	12	13
14	15	16	17	18	19	20
21	22	23	24	25	26	27
28	29	30				

   i)   1 day **before** 1 June

   ii)   2 days **after** 30 May

   iii) 1 week **before** 24 June

   iv)   1 week **before** 12 June

   v)   2 weeks **after** 8 May

   vi)   1 week **after** 25 May?

⭐ **6)** Find a calendar showing this year.
How many weeks are there from the:

   a)   first Monday in June to the last Monday in June

   b)   last Friday in May to the first Friday in June

   c)   first Monday in January to the first Monday in March

   d)   last Saturday in November to the last Saturday in December?

- -

Let's try this!

Use a blank calendar template.

Ask your classmates to tell you important dates (such as birthdays, holidays or other special days).

Mark the dates on the calendar.

Create a class display of important dates to remember.

- -

Timing events

💡 **I will learn to use the correct timer for an event.**

We use many different things to measure time.
To time an event, choose one.

clock

stopwatch

egg timer

calendar

Exercise 4

1) What would you use to time these?
Choose from **clock**, **stopwatch**, **egg timer** or **calendar**.

a) a walk around the classroom

b) break time

c) how long you sleep at night

d) how old a kitten is

e) how long to boil an egg

f) how long a maths lesson is

g) how long you can hold your breath

h) how long it takes to walk home

⭐ **2)** A stopwatch can have a 'readout' that looks like the one shown here.

a) Find out the meaning of this readout (especially the 7).

b) Investigate what sports or activities use this kind of readout.

c) Why do you think some sports or activities need to use this kind of readout?

Revisit, review, revise

1) In your jotter, write the time on each clock using words.

a)

b)

c)

d)

2) In your jotter, write the digital time for each clock.

a)

b)

c)

3) In your jotter, write the digital time for each of the following.

    a) 9 o'clock   **b)** half past ten

    c) quarter past 7   **d)** quarter to 3

4) How many days in the month of:

    a) January   **b)** February   **c)** April   **d)** June

    e) August   **f)** October   **g)** November **h)** December?

5) What is the:

    a) 6th month   **b)** 3rd month

    c) 10th month   **d)** 8th month?

6) In your jotter, write each date using words.

    a) 12/05/15   **b)** 01/08/20   **c)** 07/06/25

7) In your jotter, write each date using numbers.

    a) fifth of May two thousand and sixteen

    b) twentieth of October two thousand and nine

8) Look at the calendar page for July 2026.

2026		July			2026	
SUN	MON	TUE	WED	THU	FRI	SAT
			1	2	3	4
5	6	7	8	9	10	11
12	13	14	15	16	17	18
19	20	21	22	23	24	25
26	27	28	29	30	31	

    a) How many Fridays in July 2026?

    b) What **day** is 29 July 2026?

9) What unit of time (**seconds**, **minutes**, **hours** or **days**) would you use to measure how long it takes:

a) flying from Edinburgh to Paris

b) walking to your nearest shop

c) adding the numbers from 1 to 5

d) walking from Glasgow to London?

10) In your jotter, write an activity that would take about:

a) 10 seconds   b) 10 minutes   c) 10 hours.

11) Match each time with one of these actions:

a)	1 second
b)	2 minutes
c)	3 hours
d)	4 days
e)	15 minutes

1)	drink a pint of milk
2)	do your homework
3)	watch two films
4)	sneeze
5)	paint the whole school

12) Which measuring device would you use to time these activities?

clock

stopwatch

calendar

a) how long it takes to walk to the classroom door

b) how long it is until your next birthday

c) how long it takes to do your homework

185

15 Position and movement
More turning and directions

Full turns

 I will learn to identify full turns.

When the minute hand of a clock moves from the 12 right round to the 12 again, it makes a **full turn**. This is also known as one whole revolution.

Each time the minute hand moves to the **next number**, it moves through **5 minutes**.

Exercise 1

1) How many minutes in a:
   a) quarter turn
   b) half turn
   c) full turn?

2) How many minutes does the minute hand move through on these clock faces?

a)

b)

c)

d)

e)

f)

★ **3)** On a clock face, how many minutes does the minute hand move through when it turns clockwise from the:

a) 6 to the 9   b) 7 to the 12   c) 2 to the 5

d) 12 to the 12   e) 10 to the 2   f) 7 to the 3?

Remember, remember

The hands of a clock move in a clockwise direction.

Anticlockwise is the opposite direction.

clockwise   anticlockwise

4) Imagine a clock with hands that move anticlockwise.
How many minutes does the minute hand move through when it turns anticlockwise from the:

a) 9 to the 6 **b)** 7 to the 1 **c)** 10 to the 3

d) 3 to the 12 **e)** 5 to the 10 **f)** 8 to the 5?

Compass points

💡 **I will learn to use compass points and directions.**

The four main points of the compass are
North, South, East and **West**.

North always points to the North Pole.

Learn this rhyme to remember the clockwise order of the compass points:

Never – Eat – Smelly – Wellies.

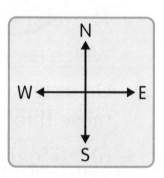

Exercise 2

 1) In your jotter, copy the compass diagram on page 191.

2) Describe the turn that is happening in each case. Choose from **quarter turn**, **half turn**, **half turn then quarter turn**.

a) East to South (clockwise)

b) North to West (anticlockwise)

c) North to West (clockwise)

d) East to South (anticlockwise)

3) Ed is walking East.
He makes a quarter turn clockwise.
Which direction is Ed walking now?

4) Jill is driving West.
She comes to a roundabout
She makes a half turn clockwise.
Which direction is Jill driving now?

5) A jet is flying South.
The jet makes a quarter turn clockwise.

Which direction is the jet travelling now?

6) A submarine is sailing West.
The submarine turns anticlockwise by a quarter turn.

Which direction is it now sailing?

7) A helicopter is facing North.
In which direction does
the helicopter need to fly
to get to:

a) the city

b) the beach

c) the airport?

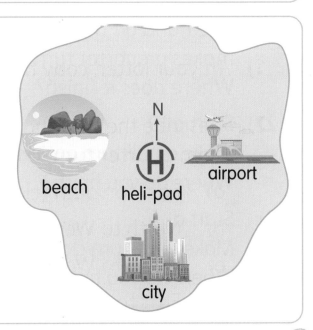

8) Look at the picture in question 7.
Ava cycles from the beach to the airport.

   a) In which direction does she travel?

   b) In which direction will her return
   journey be?

Let's try this!

Draw a map of a treasure island.

Draw a compass diagram on the map, with
North pointing to the top of the page.

Mark some places (like volcanoes, treasure, swamp, forest).

Ask a partner questions about the map using compass points.
For example: *What direction is the swamp from the volcano?*

Revisit, review, revise

1) The hand on the clock points to 6. It
makes a quarter turn anticlockwise.
Where does it finish?

2) Start at 9.
Make a half turn anticlockwise.
Where do you finish?

3) Start at 4.
Make a full turn clockwise.
Where do you finish?

4) Start at 5.
Make a full turn anticlockwise.
Where do you finish?

5) How many minutes does the clock hand move through on these clock faces?

a)

b)

c)

d)

6) How many minutes does the minute hand move through if it moves clockwise from:

a) 12 to 9   **b)** 12 to 6   **c)** 3 to 9

d) 9 to 12   **e)** 1 to 3   **f)** 4 to 8?

7) Draw a compass diagram showing North, South, East and West.

8) Jo is facing West.
He turns a quarter turn clockwise.
Which direction is he facing now?

9) Mai is facing East.
She turns a half turn anticlockwise, then a quarter turn clockwise.
Which direction is she facing now?

16 End-of-year revision

1) What number is each arrow pointing to?

a)

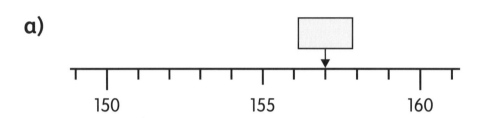

b)

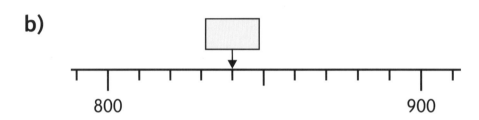

2) In your jotter, write all the **missing** numbers.

a) 418, 419, ____, ____, 422, ____

b) ____, ____, 789, 788, ____, 786

c) ____, 996, ____ , 998, ____, ____

3) a) What number is one **more** than 742?

 b) What number is one **less** than 909?

4) In your jotter, copy and complete:

529 = ____ hundreds, ____ tens and ____ ones.

5) a) In your jotter, write the number four hundred and thirty-eight using digits.

   b) In your jotter, write the number five hundred and three using digits.

6) a) In your jotter, write the number 713 in words.

   b) In your jotter, write the number 486 in words.

7) Use a method that you know to work out these additions. Show your working.

   a) 143 + 354   **b)** 127 + 212   **c)** 38 + 9

8) There are 165 cars and 13 vans in a car park.

   How many cars and vans altogether?

9) In one day, a baker sells 352 cakes.
   He also sells 216 pies.

   How many cakes and pies does he sell in total?

10) Archie is 37 years old.
   Rory is 4 years older.

   How old is Rory?

11) Use a method that you know to work out these subtractions. Show your working.

   a) 587 − 252   **b)** 72 − 8

12) There are 62 grapes in a bowl.
5 are green and the rest are red.

How many grapes are red?

13) Dan is 182 centimetres tall.
Ben is 50 centimetres shorter.

How tall is Ben?

14) What number does 72 round to, to the nearest 10?

15) What number does 430 round to, to the nearest 100?

16) What number does 150 round to, to the nearest 100?

17) What number does 384 round to:

a) to the nearest 10   **b)** to the nearest 100?

18) Copy and complete to estimate the answer to 319 + 266:

319 + 266 is about 320 + ___ = ___

19) What is:

a) double 9   **b)** treble 5?

20) In your jotter, copy and complete:

a) 2 × 4 = ___   **b)** 2 × 5 = ___

c) 2 × ___ = 6   **d)** 2 × ___ = 20

e) ___ × 2 = 14   **f)** 2 × ___ = 18

21) In your jotter, copy and complete:

a) $3 \times \underline{\quad} = 6$

b) $3 \times 1 = \underline{\quad}$

c) $\underline{\quad} \times 3 = 12$

d) $3 \times 8 = \underline{\quad}$

e) $3 \times \underline{\quad} = 30$

f) $\underline{\quad} \times 3 = 0$

22) In your jotter, copy and complete:

a) $4 \times 5 = \underline{\quad}$

b) $5 \times \underline{\quad} = 10$

c) $\underline{\quad} \times 5 = 25$

d) $5 \times 9 = \underline{\quad}$

e) $5 \times \underline{\quad} = 30$

f) $\underline{\quad} \times 5 = 50$

23) In your jotter, copy and complete:

a) $2 \times 10 = \underline{\quad}$

b) $5 \times 10 = \underline{\quad}$

c) $\underline{\quad} \times 10 = 30$

d) $10 \times 7 = \underline{\quad}$

e) $4 \times \underline{\quad} = 40$

f) $10 \times \underline{\quad} = 90$

24) In your jotter, work out:

a) $8 \div 2 = \underline{\quad}$

b) $12 \div 2 = \underline{\quad}$

c) $20 \div 10 = \underline{\quad}$

d) $70 \div 10 = \underline{\quad}$

e) $15 \div 5 = \underline{\quad}$

f) $30 \div 5 = \underline{\quad}$

g) $30 \div 3 = \underline{\quad}$

h) $9 \div 3 = \underline{\quad}$

25) In your jotter, copy and complete:

a) $2\overline{)80}$

b) $3\overline{)63}$

26) Manjinder pays £48 for 2 theatre tickets.
Each ticket costs the same.

How much does she pay for each ticket?

27) 39 bread rolls are shared equally among 3 bags.

How many rolls in each bag?

28) Flowers in a garden are planted in rows of 5.
35 flowers are planted.

How many rows are there?

29) There are 24 cans of juice in the box.

   a) How many cans in half the box?

   b) How many cans in a quarter of the box?

30) What is:

   a) half of 10   **b)** a quarter of 12?

31) How much money is there?

32) In your jotter, write each amount in two ways.
(For example: fifteen pence = 15p or £0.15)

    a)   eighty-nine pence

    b)   seventy pence

    c)   six pence

    d)   three pounds and five pence

    e)   four hundred and twenty pence

33) Freya has three £1 coins, six 10p coins and two 1p coins.

How much money does Freya have?

34) Tom buys a drink and a chocolate bar.

How much does it cost?

35) In your jotter, write the digital time for:

    a)   half past 5   **b)**   quarter past 11   **c)**   quarter to 8

    d)

    e)

    f)

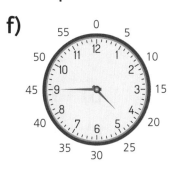

36) What number does ⬭ stand for in each calculation?

    a) $2 \times$ ⬭ $= 18$ **b)** ⬭ $\div 3 = 8$

37) In your jotter, copy and complete with **+**, **−**, **×** or **÷** to make each number sentence true.

    a)   14 ___ 2 = 7   **b)** 12 ___ 3 = 9

    c)   8 ___ 8 = 16   **d)** 5 ___ 3 = 15

38) In your jotter, write the name of each shape.

    a) 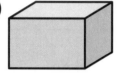   **b)**   **c)**

    d)   **e)**   **f)**

    g)   **h)**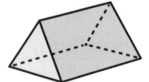

39) The top face of this cuboid is a square.

    a)   Make a list of the shapes of all the faces of this cuboid.

    b)   How many vertices does a cuboid have?

    c)   How many edges does it have?

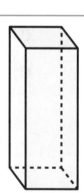

40) The minute hand of this clock points at 6.
It makes a quarter turn clockwise.

a) What number does the hand point to now?

b) How many minutes does the hand move through?

41) Where will the hand of a clock point to if:

a) it starts at 9 and moves a quarter turn clockwise

b) it starts at 4 and moves a half turn anticlockwise

c) it starts at 8 and moves a full turn anticlockwise?

42) Describe each turn (**half turn**, **quarter turn** or **half turn followed by quarter turn**).

a) South to North (clockwise)

b) South to West (clockwise)

c) South to East (anticlockwise)

d) South to West (anticlockwise)

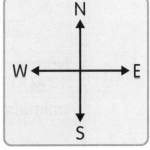

43) Do these shapes have symmetry? Write **yes** or **no**.

a)

b)

c)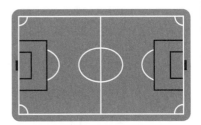

44) This pictograph shows how pupils travel to school.

Key: 人 stands for **3 pupils.**

Bike	人 人
Bus	人 人 人
Car	人 人 ⁄
Walk	人 人 人 人 ⌃

a) How many pupils travel by bike?

b) How many pupils travel by car?

c) How many more pupils walk than travel by bus?

45) This bar graph shows the number of animals on a farm.

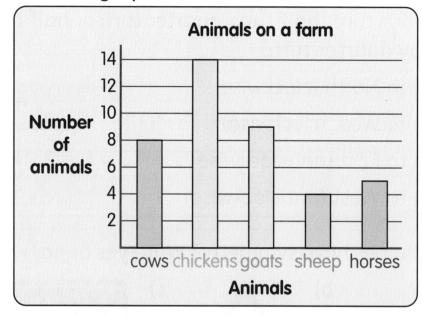

a) How many of the animals are sheep?

b) How many of the animals are goats?

c) How many more cows than horses?

46) The table shows the number of hours of sunshine in a town over five days.

Day	Monday	Tuesday	Wednesday	Thursday	Friday
Hours of sunshine	6	8	5	10	9

Copy and complete the pictograph using this key:

Key: ☀ **stands for 2 hours of sunshine.**

Monday	
Tuesday	
Wednesday	
Thursday	
Friday	

47) People were asked to name their favourite crisp flavour.

Ready salted	Salt & vinegar	Chicken	Cheese & onion
Salt & vinegar	Cheese & onion	Salt & vinegar	Ready salted
Chicken	Ready salted	Beef	Salt & vinegar
Salt & vinegar	Ready salted	Chicken	Salt & vinegar

a) In your jotter, copy and complete this table:

Favourite crisp flavour	Tally	Total
Ready salted		
Cheese and onion		
Beef		
Salt and vinegar		
Chicken		

b) Copy and complete this block graph:

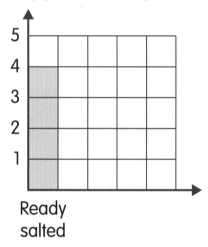

c) Copy and complete this bar graph:

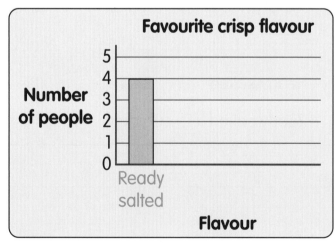

48) a) Estimate the length of this line:

_____

b) Measure the line to the nearest cm.

c) Draw a line 8 cm long.

49) In your jotter, write the unit (**mm**, **cm**, **m** or **km**) you would use to measure:

a) the thickness of a mat

b) the length of your finger

c) the distance from Edinburgh to Glasgow

d) the height of the room.

50) What would you use (**ruler**, **metre stick**, **trundle wheel** or **odometer**) to measure the things in question 49?

51) In your jotter, copy and complete:

a) 3 m = _____ cm

b) 10 cm = _____ mm

c) 2 m 5 cm = _____ cm

d) 4 cm 3 mm = _____ cm

52) In your jotter, write the area of each shape:

a)

b)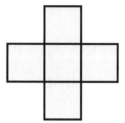

c)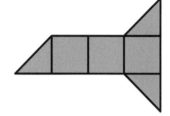

53) Estimate the area of this shape.

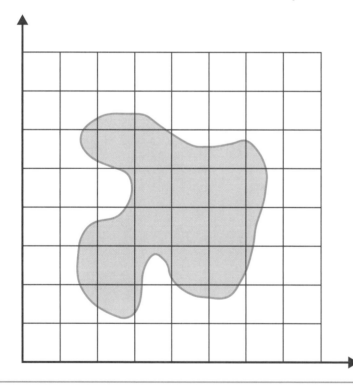

Answers

Chapter 0 (pages 8–21)

1) a) 17   b) 12
2) a) 66
   b) 30, 80, 100
3) a) 6, 8, 10, 13
   b) 33, 30, 29, 26
4) a) 7   b) 3, 2
5) a) 7   b) 90
   c) 30 + 6
6) a) 18   b) 59
7) 14
8) a) 52   b) 32
9) a) 85   b) 36
10) no
11) a) no   b) 8
12) a) no   b) yes
13) $\frac{1}{4}$
14) $\frac{1}{2}$
15) 10
16) 5
17) 10
18) 88
19) 24
20) a) 2 × 20p + 1 × 5p (or equivalent)
   b) 3 × 20p + 1 × 5p + 2 × 2p (or equivalent)

21) a) Thursday
   b) Saturday
   c) Monday
   d) Monday
22) spring
23) a) October
   b) June
   c) November
24) a) July
   b) March
25) a) half-past two
   b) quarter-to nine
   c) quarter-past eleven
26) 12:00 or 12 o'clock
27) a) 6 cm
   b) 10 cm
   c) 12 cm
28) a) 7 cm line
   b) 5 cm line
   c) 13 cm line
29) a) rectangle is 7 cm × 5 cm
   b) square is 5 cm × 5 cm
30) 1
31) a) metre stick
   b) ruler
   c) metre stick
   d) ruler

32) elephant
33) potato
34) a) 21 kg, 18 kg, 12 kg, 11 kg
   b) 99 kg, 21 kg, 17 kg, 10 kg
35) a) 3 kg
   b) $4\frac{1}{2}$ kg
36) pea, orange, watermelon
37) a)
   b)
38)
39) a) V W
   b) G F
   c) GI HJ
40) a) 17, 16
   b) 65, 60
   c) 22, 24
   d) 80, 90

Answers

41) a) 42,___,44

b) 56,___,58

c) 89,___,91

d) 38,___,40

42) 48, 49, 51

43) 52

44) a) 3 b) 8 c) 5

d) 1 e) 6 f) 3

g) 7 h) 3

45) a) 8 b) 0 c) 9

d) 2 e) 4 f) 12

g) 10 h) 15 i) 19

46) a) 3 b) 3

c) 5 d) 4

47) see pupil's drawing

48) a) 6 b) 6

c) 6 d) hexagon

49) a) any four-sided shape (square, rectangle)

b) any curved shape (circle, oval)

50) a) 4 b) 0

51) a) no b) yes c) yes

52) a) 1 b) 2

53) A

54) a) 5 b) 2 c) 2

d) 4 e) 15

55) a) Oliver

b) chicken

c) Isaac – trifle
Oliver – cheesecake

56)

	Food	Not Food
Red	A	D
Not Red	C	B, E

57) a) rectangle

b) square

c) circle

58) a) C6 b) F1 c) B2

59) forward 2
turn right
forward 2
turn left
forward 3
turn left
forward 3
turn right
forward 2

Chapter 1 Whole numbers 1

Exercise 1 (page 22)

1) a) 3, 7 b) 4, 8

2) a) 6, 5, 7 b) 9, 0, 2

c) 7, 6, 0

3) a) ones

b) tens, ones

c) hundreds, tens, ones

d) hundreds, tens, ones

4) a) 3 hundreds, 0 tens and 0 ones

b) 6 hundreds, 0 tens and 2 ones

c) 2 hundreds, 7 tens and 5 ones

5) a) 3

b) 30, 8

c) 300, 20

d) 300, 2

6) a) tens

b) ones

c) hundreds

Exercise 2 (page 25)

1) a) 169, 171, 172

b) 418, 420, 421, 422

c) 996, 997, 999, 1000

2) a) 652, 651, 648

b) 220, 219, 217, 216

c) 999, 998, 996, 995

3) a) 411 b) 167

c) 790 d) 1000

4) a) 627 b) 561

c) 930 d) 869

5) 103

6) a) 260 b) 790

7) 400

8) 439p

Exercise 3 (page 27)

1) a) 63 b) 136

c) 163 d) 336

e) 103

2) a) 23 b) 38 c) 41

d) 75 e) 62 f) 90

g) 12 h) 80

3) a) 243 b) 714

c) 970 d) 808

4) a) fifty

b) thirteen

c) nine hundred

d) six hundred, one

e) seven hundred, thirty

5) a) fifty-eight

b) forty-six

c) nineteen

d) seventy

e) eight-eight

f) three hundred and eleven

g) five hundred and four

h) one thousand

Revisit, review, revise (page 29)

1) a) 5, 1, 2

b) hundreds, tens, 8

c) 40 d) 1

2) tens

3) a) 89 b) 37

4) a) 460, 459

b) 129, 130, 132

5) a) 386

b) 800

c) 440

6) a) i) 63 ii) 162

b) i) seventy-eight

ii) four hundred and nine

iii) seven hundred and fifty-three

iv) one thousand

Chapter 2: Symmetry

Exercise 1 (page 31)

1) all pictures symmetrical

2) a) yes b) no

c) yes d) no

e) no f) yes

g) yes h) no

i) yes j) yes

k) no l) yes

3) a) yes

b) yes

c) no

d) yes

e) no

f) yes

4) see pupil's answer

5) see pupil's answer

Revisit, review, revise (page 33)

1) see pupil's answer

2) see pupil's answer

Chapter 3 Whole numbers 2

Exercise 1 (page 35)

1) a) 4

b) 200, 80

c)
```
    6  1  4
 +  2  8  5
 ─────────
    8  9  9
```

2) a) 227 b) 639

c) 449 d) 897

e) 797 f) 749

g) 679 h) 793

i) 699 j) 949

k) 987 l) 987

3) 9, 9

4) a) 500 b) 900

c) 412 d) 873

e) 360 f) 917

5) a) 689 b) 958

c) 886 d) 438

e) 199 f) 239

6) 4 hundreds + 5 tens = 450

7) a) 630 b) 520

c) 290 d) 383

e) 475 f) 380

8) £580

9) 692

10) 699

11) a) 283 b) no

Answers

Exercise 2 (page 39)

1) a) 52 b) 31 c) 74
   d) 84 e) 94 f) 48
   g) 45 h) 46 i) 37

2) a) 46 b) 31 c) 83
   d) 92 e) 95 f) 40
   g) 65 h) 86

3) 34p

4) 55p

5) 71

6) 48

7) 63

8) £30

Revisit, review, revise (page 42)

1) a) 61   b) 25
   c) 73   d) 84

2) a) 898   b) 799
   c) 397   d) 909
   e) 754   f) 179

3) a) 700   b) 571
   c) 650   d) 973

4) 338

5) £35

6) 349

7) 897

Chapter 4 Money

Exercise 1 (page 43)

1) a) £0.95 b) £0.36
   c) £0.20 d) £0.13

   e) £0.99 f) £1.00
   g) £0.03 h) £0.09
   i) £0.02

2) a) 45p   b) 72p
   c) 83p   d) 21p
   e) 50p   f) 75p
   g) 100p h) 4p
   i) 5p

3) a) £0.71 or 71p
   b) £0.22 or 22p
   c) £0.60 or 60p
   d) £0.06 or 6p

Exercise 2 (page 45)

1) a) £4.60 b) £4.46
   c) £4.71 d) £3.08

2) a) £1.94 b) £2.30
   c) £4.01 d) £4.99
   e) £3.17 f) £1.09

3) a) 412p   b) 210p
   c) 301p   d) 450p
   e) 300p   f) 99p

4) a) £1.20 b) £3.00
   c) £2.36 d) £5.00
   e) £3.34

5) Len

6) Answers may vary, i.e. £2, £2, 20p, 10p, 1p

Exercise 3 (page 50)

1) a) £2.15 b) £1.25
   c) £3.40

2) a) £0.80 b) £1.40
   c) £1.90 d) £2.10

3) a) £8.17 b) £1.83

4) £3.87

5) £3.98

Revisit, review, revise (page 51)

1) a) £0.62 b) £0.09
   c) £0.51 d) £0.90
   e) £2.10 f) £3.09

2) a) £1.95 b) £1.39
   c) £7.57

3) a) £2.57 b) £6.01

4) £2.95

5) £2.17

Chapter 5 Whole numbers 3

Exercise 1 (page 54)

1) a) 8   b) 200, 30
   c)
   $$\begin{array}{r} 6\ 7\ 8 \\ -\ 2\ 3\ 7 \\ \hline 4\ 4\ 1 \end{array}$$

2) a) 163   b) 204
   c) 304   d) 142
   e) 312   f) 351
   g) 315   h) 272
   i) 532   j) 333
   k) 21   l) 1

3) 3, 3

4) a) 500   b) 300

   c) 200   d) 712

   e) 200   f) 400

5) a) 111   b) 751

   c) 531   d) 611

   e) 814   f) 282

6) 3, 2, 320

7) a) 500   b) 240

   c) 910   d) 830

   e) 320   f) 600

8) 171

9) 310

10) a) £279   b) £104

Exercise 2 (page 58)

1) a) 18   b) 26

   c) 19

2) a) 39   b) 23

   c) 59

3) a) 39   b) 78

   c) 58   d) 28

   e) 68   f) 44

   g) 68   h) 85

   i) 16

4) 17

5) 58

6) 80

Revisit, review, revise (page 60)

1) a) 54   b) 36

   c) 79

2) a) 68   b) 56

   c) 23   d) 17

3) a) 233   b) 342

   c) 661

4) a) 300   b) 529

   c) 200   d) 700

   e) 420   f) 200

5) 114

6) £472

Chapter 6 Whole numbers 4

Exercise 1 (page 61)

1) a) 70   b) 80

   c) 80   d) 70, 80

   e) 70

2) a) 130

   b) 130, 120

   c) 130

3) a) 30

   b) 20, 30

   c) 20

4) a) 90, 90

   b) 130, 130

   c) 460, 460

   d) 910, 900

5) a) 40   b) 70

   c) 20   d) 60

   e) 160   f) 140

   g) 320   h) 530

   i) 50

6) a) 600

   b) 700

   c) 600, 700

   d) 600, 700

   e) 600

7) a) 700   b) 700

8) a) 2, 3, 300

   b) 8, 9, 800

   c) 3, 4, 300

9) a) 100   b) 400

   c) 700   d) 400

   e) 900   f) 600

   g) 200   h) 800

10) a) 30 cm   b) £200

   c) 630

Exercise 2 (page 65)

1) a) 90   b) 40

   c) 90 – 40 = 50

2) a) 100

   b) 80, 100

   c) 20, 70

   d) 30, 50, 80

   e) 150, 360

   f) 730, 150, 880

3) a) 20, 50

   b) 90, 80, 10

   c) 30, 120

Answers

d) 140, 50

e) 450, 200, 250

f) 580, 220, 360

4) a) 90   b) 150

   c) 100   d) 270

   e) 40   f) 150

5) £240

Revisit, review, revise (page 68)

1) 70

2) a) 40   b) 70

   c) 460   d) 290

3) 400

4) a) 800   b) 400

   c) 800   d) 500

5) 150, 400

6) a) 70   b) 660

7) 120

8) 220

Chapter 7 3D shapes

Exercise 1 (page 70)

1) a) cube

   b) cylinder

   c) sphere

d) cuboid

e) cone

2) a) cone

   b) cube

   c) cylinder

   d) sphere

   e) cuboid

3) a) hemisphere

   b) triangular prism

   c) square-based prism

4) a) square-based pyramid

   b) cube

   c) cone

   d) cuboid

   e) sphere

   f) cylinder

   g) cone

   h) triangular prism

5) a) cuboid, triangular prism

   b) cone, hemisphere

   c) cylinder, cone

   d) cone, cylinder, hemisphere

   e) cube, cylinder, cuboid

   f) cube, square-based pyramid

   g) cube, cuboid, square-based pyramid, cone

h) cylinders (3), sphere

i) cylinders (4), cuboid, triangular prism

Exercise 2 (page 74)

1) rectangles

2) rectangles and squares

3) a) triangles and a square

   b) triangles and rectangles

4) a) cone

   c) sphere

   d) hemisphere

   e) cylinder

5) a) cube

   b) cuboid

   c) triangular prism

   d) square-based pyramid

   e) cone

   f) cylinder

Exercise 3 (page 76)

1) a) 6   b) 8   c) 12

2) a) 6   b) 8   c) 12

3) a) 5   b) 5   c) 8

4) a) 9   b) 6   c) 5

5) a) 1   b) 1   c) 2

6) a) 1   b) 0   c) 0

7) a) 0   b) 2   c) 3

8)

	Faces	Vertices	Edges
Cuboid	6	8	12
Cube	6	8	12
Square-based pyramid	5	5	8
Triangular prism	5	6	9
Cone	2	1	1
Sphere	1	0	0
Cylinder	3	0	2

9) cone, sphere, cylinder

Revisit, review, revise (page 78)

1) a) cube
   b) sphere
   c) square-based pyramid
   d) cone
   e) cylinder
   f) cuboid
   g) triangular prism

2 1 sphere
   1 cylinder
   1 cone
   1 cube
   1 triangular prism
   3 cuboids

3) triangles and a square

4) a) cube
   b) triangular prism
   c) cuboid
   d) square-based pyramid

5) a) 4   b) 8   c) 12
6) a) 12   b) 8
   c) 8   d) 6

Chapter 8 Whole numbers 5

Exercise 1 (page 82)

1) 6
2) 10
3) a) 4   b) 20   c) 12
4) a) 10   b) 20

Exercise 2 (page 84)

1) 6
2) 15
3) a) 9   b) 30   c) 18
4) 24

Exercise 3 (page 85)

1) a) 10   b) 12   c) 14
   d) 16   e) 18   f) 20
2) 4, 6, 8, 10, 12, 14, 16, 18, 20
3) see pupil's answer
4) 8, 10, 12, 14, 16, 18, 20
5) pupil's own practice
6) a) 2   b) 4   c) 6
   d) 9   e) 7   f) 8
   g) 5   h) 10   i) 0
7) 8
8) 14
9) 16

10) a) 6   b) 20
    c) 6, 2, 12 d) 5, 5
    e) 9, 9, 2

Exercise 4 (page 90)

1) a) 15   b) 18   c) 21
   d) 24   e) 27   f) 30
2) 3, 6, 9, 12, 15, 18, 21, 24, 27, 30
3) see pupil's answer
4) 12, 15, 18, 21, 24, 27, 30
5) pupil's own practice
6) a) 3   b) 4   c) 7
   d) 5   e) 8   f) 10
   g) 9   h) 6   i) 0
7) 15
8) 6 m
9) 21
10) a) 6   b) 15
    c) 6, 3, 18 d) 10, 10
    e) 3, 3, 3

Revisit, review, revise (page 94)

1) a) 6   b) 12   c) 4
   d) 10   e) 2   f) 5
   g) 1   h) 7   i) 18
   j) 8   k) 0
2) a) 30   b) 6   c) 4
   d) 10   e) 3   f) 1
   g) 5   h) 6   i) 24
   j) 9   k) 0

Answers

3) a) 7 × 2 or 2 × 7

   b) 9 × 2 or 2 × 9

   c) 2 × 3 or 3 × 2

   d) 5 × 3 or 3 × 5

4) a) 4   b) 16

   c) 9   d) 30

Chapter 9 Measurement

Exercise 1 (page 96)

1) see pupil's answer

2) a) 5 cm   b) 7 cm

   c) 2 cm   d) 8 cm

   e) 3 cm   f) 9 cm

3) a) f, d, b, a, e, c

   b) 7 cm

4) see pupil's answer

5) a) 4 cm   b) 3 cm

   c) 7 cm   d) 3 cm

   e) 5 cm

6) a) 10 cm, 4 cm, 9 cm and 2 cm

   b) 10 cm – 2 cm = 8 cm

Exercise 2 (page 98)

1) a) 100 cm

   b) 500 cm

   c) 800 cm

   d) 300 cm

   e) 900 cm

   f) 1000 cm

2) a) 10 mm

   b) 70 mm

   c) 80 mm

   d) 100 mm

   e) 90 mm

   f) 40 mm

3) a) 4 m   b) 7 m

   c) 8 m   d) 10 m

   e) 2 m   f) 1 m

4) a) 2 cm

   b) 10 cm

   c) 8 cm

5) a) 64 mm

   b) 93 mm

   c) 31 mm

   d) 79 mm

6) a) 210 cm

   b) 925 cm

   c) 318 cm

   d) 180 cm

   e) 345 cm

   f) 503 cm

Exercise 3 (page 102)

1) a) cm   b) km

   c) m   d) mm

   e) cm   f) mm

   g) km   h) m

2) a) ruler

   b) ruler

   c) metre stick/tape, trundle wheel

   d) odometer

3) see pupil's answer

4) a) trundle wheel

   b) It would take too long with a ruler.

Exercise 4 (page 104)

1) 10 cm²

2) a) 4 cm²

   b) 12 cm²

   c) 15 cm²

   d) 10 cm²

   e) 6 cm²

   f) 9 cm²

   g) 6 cm²

   h) 8 cm²

   i) 11 cm²

3) rectangles 2 × 10 or 4 × 5

4) a) 16 cm²

   b) 16 cm²

   c) They have the same area.

5) see pupil's answer

6) a) $3\frac{1}{2}$ cm²

   b) $4\frac{1}{2}$ cm²

c) 4 cm²

d) 12 cm²

e) 12 cm²

f) 12 cm²

g) 24 cm²

h) 16 cm²

Exercise 5 (page 107)

1) 16 cm²

2) a) 22 cm²

b) 24 cm²

c) 32 cm²

d) 43 cm²

Revisit, review, revise (page 108)

1) a) see pupil's answer

b) A 14 cm, B 8 cm, C 14 cm, D 11 cm

c) B, D, A, C or B, D, C, A

2) a) m   b) mm

c) cm   d) km

3) a) trundle wheel/metre stick or tape

b) ruler

c) tape measure/metre stick/ruler

d) odometer

4) a) 200   b) 40

c) 117   d) 340

e) 308

5) a) 8 cm × 3 cm rectangle

b) 7 cm × 7 cm square

c) 5 cm × 2.5 cm rectangle

6) a) this page

b) the floor

7) a) 6 cm²

b) 10 cm²

c) 10 cm²

d) 8 cm²

e) 7 cm²

f) 10 cm²

8) a) 18 cm²

b) 21 cm²

Chapter 10 Whole numbers 6

Exercise 1 (page 113)

1) a) pupil is to count

b) 4   c) 4, 4

2) a) pupil is to count

b) 5   c) 5, 5

3) a) pupil is to count and share

b) 7, 7

4) a) 6   b) 8

5) a) 5   b) 9

6) a) 1   b) 2   c) 6

7) a) 3, 3, 3   b) 5, 5, 5

8) a) 10, 10, 10

b) 6, 6, 6

c) 8, 8, 8

d) 7, 7, 7

9) a) 2 × 3 = 6
3 × 2 = 6
6 ÷ 2 = 3

b) 2 × 9 = 18
9 × 2 = 18
18 ÷ 2 = 9

Exercise 2 (page 116)

1) a) 4   b) 3   c) 5

d) 7   e) 9   f) 8

g) 10   h) 1   i) 0

2) a) 6   b) 12   c) 18

d) 16   e) 14   f) 20

3) 4

4) 6

5) 7

6) 3

7) 10

Exercise 3 (page 120)

1) a) 1   b) 4   c) 14

2) a) 13   b) 23   c) 11

d) 22   e) 31   f) 34

g) 42   h) 41

3) a) 32   b) 44   c) 21

4) 23

5) 40

6) 20

7) 24

Answers

Exercise 4 (page 123)

1) a) pupil is to count
   b) 4
   c) 4, 4
2) a) pupil is to count
   b) 5
   c) 5, 5
3) a) pupil is to count and share
   b) 10, 10
4) 8
5) a) 5   b) 7
6) a) 3 b) 2 c) 6
7) a) 2, 2, 2,
   b) 5, 5, 5
8) a) 10, 10, 10
   b) 6, 6, 6
   c) 8, 8, 8
   d) 3, 3, 3
9) a) 3 × 7 or 7 × 3
   b) 3 × 9 or 9 × 3

Exercise 5 (page 126)

1) a) 1 b) 5 c) 4
   d) 8 e) 7 f) 9
   g) 10 h) 0 i) 6
2) a) 21 b) 9 c) 18
   d) 15 e) 24 f) 30
3) 2
4) 9
5) 4
6) 10

---

7) a) 13   b) 22
   c) 10   d) 23
8) a) 12   b) 30
   c) 32   d) 31
   e) 20   f) 11
9) 12
10) 33
11) 32

Exercise 6 (page 130)

1) a) 4 b) 5 c) 10
   d) 10 e) 9 f) 8
2) a) 18 b) 18 c) 30
   d) 14 e) 12 f) 28
3) a) 2 b) 10 c) 2
   d) 3 e) 3 f) 7
   g) 15 h) 20

Exercise 7 (page 131)

1) a) +   b) -
2) a) ×   b) ÷
3) a) ×   b) -
   c) ÷   d) +
   e) ÷   f) +
4) a) -   b) +
   c) ×   d) ÷
   e) -   f) +
   g) ÷   h) +
   i) ×   j) ÷
5) a) × 5 or + 16
   b) + 16 or × 5

Revisit, review, revise (page 133)

1) a) 5   b) 7   c) 9
   d) 6   e) 9   f) 7
2) a) 23   b) 22
3) a) 32   b) 33
   c) 31   d) 32
4) 10
5) £5
6) 42
7) 21
8) a) 3   b) 4
   c) 10   d) 9
   e) 12   f) 9
   g) 30   h) 0
9) a) -   b) +
   c) ×   d) ÷
   e) +   f) -
   g) ×   h) ×
   i) ÷   j) -
   k) +   l) ÷
10) 8 × 2 = 13 + 3
    24 − 14 = 30 ÷ 3
    10 + 11 = 3 × 7

Chapter 11 Fractions

Exercise 1 (page 135)

1) 2
2) 10
3) 8
4) a) 3 b) 4 c) 7
5) a) 5 b) 9 c) 50

6) a) 25

   b) 150

   c) 230

Exercise 2 (page 137)

1) a) 2   **b)** 1

2) 2

3) 6

4) 7

5) a) 10   **b)** 25

Revisit, review, revise (page 139)

1) a) 8   **b)** 4

2) a) 12   **b)** 6

3) a) 10   **b)** 5

4) 7

Chapter 12 Statistics

Exercise 1 (page 141)

1) a) 2   **b)** 6

   c) 7   **d)** 9

   e) 10

2) a) 7 **b)** 5 **c)** 25

   d) Friday; no-one brought a packed lunch.

3) a) 9   **b)** 11

   c) November

   d) January

   e) 18

Exercise 2 (page 143)

1)

Cod	🐟🐟🐟🐟
Haddock	🐟🐟🐟🐟🐟🐟
Plaice	🐟🐟🐟🐟
Bass	🐟🐟
Sole	🐟🐟🐟🐟🐟

2)

Lion	🙂🙂🙂🙂🙂
Tiger	🙂🙂🙂
Giraffe	🙂🙂🙂🙂
Seal	🙂
Panda	🙂🙂🙂🙂🙂🙂

3)

Mon	🐦🐦
Tues	🐦🐦🐦🐦🐦
Wed	🐦🐦🐦
Thu	🐦🐦🐦🐦
Fri	🐦

4)

6:30	☥☥☥☥☥
7:00	☥☥☥☥☥☥
7:30	☥☥☥☥
8:00	☥☥⸕
8:30	☥⸕

Exercise 3 (page 145)

1) a) 5   **b)** 1

   c) 2   **d)** 10

2) a) 3   **b)** 6

   c) 13   **d)** 5

3) a) 5   **b)** 3

   c) 7   **d)** 0

   e) Duty Calls

   f) 19

4) a) 2s

   b) cupcake

   c) 10   **d)** 5

   e) 11   **f)** 3

5) a) 10   **b)** 17

   c) holiday, 22

   d) 4

6) a) 4   **b)** 22

   c) 7   **d)** 5

Exercise 4 (page 149)

1)

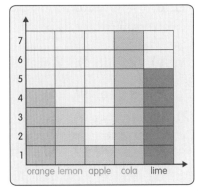

2) a)

215

Answers

b)

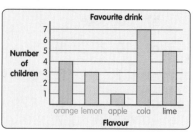
Favourite drink bar chart — Number of children vs Flavour (orange, lemon, apple, cola, lime)

3)

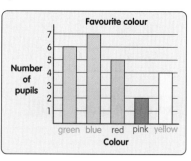
Favourite colour bar chart — Number of pupils vs Colour (green, blue, red, pink, yellow)

4)

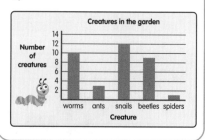
Creatures in the garden bar chart — Number of creatures vs Creature (worms, ants, snails, beetles, spiders)

Exercise 5 (page 152)

1) a) 4   b) 3

    c) 1   d) blue

    e) 2   f) 2

    g) 8   h) 15

2) a)

Town	Tally	Total
Ayr	|||	3
Airdrie	ⅢⅠ |	6
Annan	|	1

    b) 10

---

3)

City	Tally	Total
Dundee	ⅢⅠ	5
Inverness	|	1
Glasgow	ⅢⅠ |||	8
Edinburgh	ⅢⅠ ||	7
Stirling	||	2
Perth	||||	4

4) a)

City	Tally	Total
Orange	|||	3
Red	ⅢⅠ |||	8
Black	|||	3
Blue	|	1
Pink	ⅢⅠ	5

    b) red   c) 20

    d) 17

5) a) 8   b) 12   c) 19

    d) 26   e) 37

6) a) ⅢⅠ \|\|

    b) ⅢⅠ ⅢⅠ

    c) ⅢⅠ ⅢⅠ ⅢⅠ

    d) ⅢⅠ ⅢⅠ ⅢⅠ \|\|

    e) ⅢⅠ ⅢⅠ ⅢⅠ ⅢⅠ

    f) ⅢⅠ ⅢⅠ ⅢⅠ ⅢⅠ ⅢⅠ \|

---

7)

Flower	Tally	Total
Rose	ⅢⅠ	5
Tulip	ⅢⅠ ⅢⅠ	10
Pansy	ⅢⅠ ||	7
Daffodil	ⅢⅠ |	6
Marigold	||	2

Revisit, review, revise (page 155)

1) a) 8   b) 5

    c) 4   d) 1

2) a)

Order	Tally	Total
Salad	ⅢⅠ ||	7
Soup	||||	4
Omelette	ⅢⅠ ⅢⅠ	10
Burger	||	2

b)

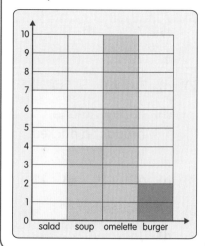
Bar chart — salad, soup, omelette, burger

c)

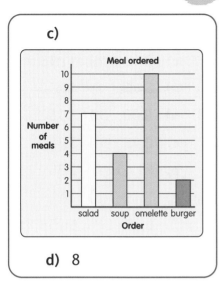

d) 8

Chapter 13 Whole numbers 7

Exercise 1 (page 159)

1) b) 10   c) 15
   d) 20   e) 25
   f) 30   g) 7, 35
   h) 8, 40   i) 9, 45
   j) 5, 50

2) a) 5, 10, 15, 20, 25, 30, 35, 40, 45, 50
   b) 0 or 5

3) 20, 25, 30, 35, 40, 45, 50

4) pupil's own practice

5) a) 5   b) 1
   c) 3   d) 7
   e) 9   f) 4
   g) 6   h) 8
   i) 0

6) 35

7) 45p

8) a) 10   b) 15
   c) 30

Exercise 2 (page 162)

1) a) 3   b) 3, 3
2) a) 5, 1
   b) 10, 2
   c) 15, 3
3) 5 × 1 = 5
   5 × 2 = 10
   5 × 3 = 15
   5 × 4 = 20
   5 × 5 = 25
   5 × 6 = 30
   5 × 7 = 35
   5 × 8 = 40
   5 × 9 = 45
   5 × 10 = 50
4) a) 4   b) 2
   c) 10   d) 7
   e) 25   f) 8
   g) 9   h) 6
5) a) 15   b) 35
   c) 50   d) 30
   e) 40   f) 25
6) a) 9   b) 4
   c) 8   d) 7
   e) 6   f) 5
   g) 10p

Exercise 3 (page 165)

1) a) 10   b) 20
   c) 30   d) 40
2) a) 50   b) 60
   c) 7, 70   d) 8, 80

e) 10, 9, 90
f) 10, 100
3) a) 10, 20, 30, 40, 50, 60, 70, 80, 90, 100
   b) 0
4) 30, 40, 50, 60, 70, 80, 90, 100
5) pupil's own practice
6) a) 5   b) 2
   c) 8   d) 3
   e) 6   f) 7
   g) 9   h) 4
   i) 10   j) 0
7) 70p
8) 80
9) 50
10) a) 20   b) 40
    c) 60   d) 100

Exercise 4 (page 168)

1) a) pupil is to count
   b) 3   c) 3, 3
2) a) 10, 1 b) 20, 2
   c) 30, 3
3) 10 × 1 = 10
   10 × 2 = 20
   10 × 3 = 30
   10 × 4 = 40
   10 × 5 = 50
   10 × 6 = 60

Answers

10 × 7 = 70
10 × 8 = 80
10 × 9 = 90
10 × 10 = 100

4) a) 5   b) 4
c) 8   d) 7
e) 6   f) 9

5) a) 30   b) 50
c) 90   d) 80
e) 60   f) 100

6) a) i) 7 ii) 5
b) 8   c) 6
d) 3   e) 10

7) 4

Revisit, review, revise (page 170)

1) a) 40   b) 15
c) 30   d) 25
e) 35   f) 5
g) 45   h) 0
i) 10   j) 20
k) 50   l) 15

2) a) 70   b) 90
c) 30   d) 60
e) 20   f) 40
g) 10   h) 0
i) 80   j) 100
k) 50   l) 80

3) a) 7   b) 9
c) 8   d) 9
e) 3   f) 4
g) 2   h) 8

4) a) 2, 2, 2
b) 3, 3, 3

5) a) 8, 8, 8
b) 10, 10, 10

6) 70

7) 9

8) a) £50
b) £20
c) £35

Chapter 14 Time
Exercise 1 (page 174)

1) a) half past 4
b) quarter past 5
c) quarter to 2
d) half past 8
e) 9 o'clock
f) quarter to 7
g) half past 2
h) quarter to 10
i) quarter past 11
j) half past 9
k) quarter past 1
l) quarter to 6

2) a) 5:15   b) 7:30
c) 2:45   d) 11:15
e) 8:45   f) 10:30

3) a) 9:30   b) 1:15
c) 8:45   d) 3:15
e) 11:45   f) 2:30
g) 7:15   h) 10:45
i) 6:30   j) 5:45

Exercise 2 (page 176)

1) a) seconds
b) days
c) hours
d) seconds
e) minutes
f) hours

2) see pupil's answer

3) more

4) more

5) less

6) read out the alphabet – 15 seconds
watch a film – 2 hours
walk a mile – 20 minutes

7) drive from Dundee to Inverness – 2 hours
grow a small plant – 5 weeks
eat Breakfast – 10 minutes
write your names – 8 seconds

8) a) 20–30 seconds

    b) 9–10 minutes

    c) 2–5 hours

9) see pupil's answer

Exercise 3 (page 179)

1) a) 23.02.23

    b) 19.04.24

    c) 22.07.19

    d) 18.08.22

    e) 07.06.14

    f) 03.03.30

    g) 10.12.16

    h) 04.02.21

2) a) 14 January 2024

    b) 1 March 2025

    c) 11 November 2016

    d) 24 April 2011

    e) 12 December 2017

    f) 07 August 2020

    g) 09 March 2025

    h) 30 June 2021

    i) 28 February 2015

3) There are not 30 days in February.

4) a) 5

    b) Thursday

    c) 24 April 2025

    d) Sunday

5) a) 9

    b) 4 June 2026

    c) i) 30 May 2026

       ii) 2 June 2026

       iii) 17 June 2026

       iv) 5 June 2026

       v) 22 May 2026

       vi) 1 June 2026

6) a) 3 or 4 depending on year

    b) 1

    c) 8 or 9 depending on year

    d) 4 or 5 depending on year

Exercise 4 (page 182)

1) a) stopwatch

    b) clock

    c) clock

    d) calendar

    e) egg timer

    f) clock

    g) stopwatch

    h) clock/stopwatch

2) a) 3 minutes, 29 seconds and 7 tenths of a second

    b) see pupil's answer

    c) In order to be very accurate, sometimes the difference in time can be very small.

Revisit, review, revise (page 183)

1) a) 5 o'clock

    b) half past 8

    c) quarter past 2

    d) quarter to 12

2) a) 4:00   b) 9:30

    c) 1:45

3) a) 9:00   b) 10:30

    c) 7:15   d) 2:45

4) a) 31   b) 28/29

    c) 30   d) 30

    e) 31   f) 31

    g) 30   h) 31

5) a) June

    b) March

    c) October

    d) August

6) a) 12 May 2015

    c) 1 August 2020

    d) 7 June 2005

Answers

7) a) 05.05.16

b) 20.10.09

8) a) 5

b) Wednesday

9) a) hours

b) minutes

c) seconds

d) days

10) see pupil's answer

11) a) 4 b) 1 c) 3

d) 5 e) 2

12) a) stopwatch

b) calendar

c) clock

Chapter 15 Position and movement

Exercise 1 (page 186)

1) a) 15 b) 30 c) 60

2) a) 15 b) 35 c) 30

d) 30 e) 50 f) 25

3) a) 15 b) 25 c) 15

d) 60 e) 20 f) 40

4) a) 15 b) 30 c) 35

d) 15 e) 35 f) 15

Exercise 2 (page 188)

1) see pupil's diagram

2) a) quarter turn

b) quarter turn

c) half turn then quarter turn

d) half turn then quarter turn

3) South

4) East

5) West

6) South

7) a) South b) West

c) East

8) a) West

b) East

Revisit, review, revise (page 190)

1) 3

2) 3

3) 4

4) 5

5) a) 10 b) 20

c) 25 d) 35

6) a) 45 b) 30

c) 30 d) 15

e) 10 f) 20

7) see pupil's diagram

8) South

9) North

Chapter 16 End-of-year revision (pages 192–204)

1) a) 157 b) 840

2) a) 420, 421, 423

b) 791, 790, 787

c) 995, 997, 999, 1000

3) a) 743 b) 908

4) 5, 2, 9

5) a) 438 b) 503

6) a) seven hundred and thirteen

b) four hundred and eighty-six

7) a) 497

b) 339

c) 47

8) 198

9) 562

10) 41

11) a) 335 b) 64

12) 57

13) 132

14) 70

15) 400

16) 200

17) a) 380 b) 400

18) 270, 590

19) a) 18 b) 15

20) a) 8 b) 10 c) 3

d) 10 e) 7 f) 9

21) a) 2 b) 3 c) 4

d) 24 e) 10 f) 0

22) a) 20 b) 2 c) 5

d) 45 e) 6 f) 10

23) a) 20 b) 50 c) 3

d) 70 e) 10 f) 9

24) a) 4 **b)** 6 **c)** 2
d) 7 **e)** 3 **f)** 6
g) 10 **h)** 3

25) a) 40 **b)** 21

26) £24

27) 13

28) 7

29) a) 12 **b)** 6

30) a) 5 **b)** 3

31) £3.77

32) a) 89p, £0.89
b) 70p, £0.70
c) 6p, £0.06
d) 305, £3.05
e) 420, £4.20

33) £3.62

34) £2.55

35) a) 5:30 **b)** 11:15
c) 7:45 **d)** 8:00
e) 3:30 **f)** 4:45

36) a) 9 **b)** 24

37) a) ÷ **b)** –
c) + **d)** ×

38) a) cuboid
b) sphere
c) cylinder
d) cube
e) cone
f) square-based pyramid
g) hemisphere
h) triangular prism

39) a) rectangles, squares
b) 8 **c)** 12

40) a) 9 **b)** 15

41) a) 12 **b)** 10 **c)** 8

42) a) half turn
b) quarter turn
c) quarter turn
d) half turn followed by quarter turn

43) a) yes **b)** no
c) yes

44) a) 6 **b)** 7 **c)** 5

45) a) 4 **b)** 9 **c)** 3

46)

Monday	☀ ☀ ☀
Tuesday	☀ ☀ ☀ ☀
Wednesday	☀ ☀ ☀
Thursday	☀ ☀ ☀ ☀ ☀
Friday	☀ ☀ ☀ ☀ ☀

47) a)

Favourite crisp flavour	Tally	Total
Ready salted	||||	4
Cheese and onion	||	2
Beef	|	1
Salt and vinegar	ⅢⅢ|	6
Chicken	|||	3

b)

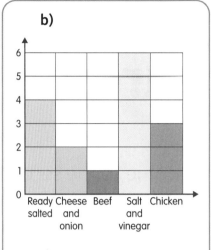

c)

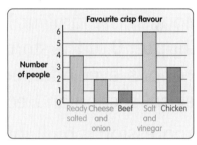

48) a) see pupil's answer
b) 9 cm
c) pupil's straight line of 8 cm

49) a) mm **b)** cm
c) km **d)** m

50) a) ruler **b)** ruler
c) odometer
d) metre stick

51) a) 300 **b)** 100
c) 205 **d)** 403

52) a) 8 cm² **b)** 5 cm²
c) $4\frac{1}{2}$ cm²

53) 18 cm²

Information for teachers, parents and carers

Welcome to the second edition of a well-loved TeeJay Maths series. The First Level scheme has been restructured so that it comprises three books instead of two, in line with curriculum structure. Book 1A covers the course for P2, Book 1B covers P3 and Book 1C covers P4.

Many of the familiar TeeJay features have been retained, including a **Chapter 0** at the start of each book, which revisits topics learned at the previous level. Additionally, each chapter ends with a '**Revisit, review, revise**' section and, each book ends with an **End-of-year revision** chapter.

Progression is built into the structure of each book, with Whole Number chapters alternating with other topics. Questions for differentiation have been flagged throughout:

- Easier questions/activities, or building blocks, are flagged by this icon.
- Hard questions, or stretch, are flagged by this star icon.

Activities for **play-based learning** (Let's try this!) have been embedded throughout to engage pupils in their learning.

Answers to all questions can be found at the back of each book for ease of use.

In addition to the three second edition textbooks, new **interactive resources**, **editable course plans**, **teaching guides** and **worksheets** will be available through our Boost platform. The worksheets include **mental maths**, **practice** (to test pupils on each unit) and **assessment** (to be taken at half-term). The practice and assessment worksheets are available in digital and PDF format. The mental maths are PDF only.

http://hoddergibson.co.uk/teejayfirstlevelmaths